Lecture Notes in Applied and Computational Mechanics

Volume 47

Series Editors

Prof. Dr.-Ing. Friedrich Pfeiffer
Prof. Dr.-Ing. Peter Wriggers

Lecture Notes in Applied and Computational Mechanics

Edited by F. Pfeiffer and P. Wriggers

Further volumes of this series found on our homepage: springer.com

Vol. 47: Studer, C.
Numerics of Unilateral Contacts and Friction
191 p. 2009 [978-3-642-01099-6]

Vol. 46: Ganghoffer, J.-F., Pastrone, F. (Eds.)
Mechanics of Microstructured Solids
136 p. 2009 [978-3-642-00910-5]

Vol. 45: Shevchuk, I.V.
Convective Heat and Mass Transfer in Rotating Disk Systems
300 p. 2009 [978-3-642-00717-0]

Vol. 44: Ibrahim R.A., Babitsky, V.I., Okuma, M. (Eds.)
Vibro-Impact Dynamics of Ocean Systems and Related Problems
280 p. 2009 [978-3-642-00628-9]

Vol.43: Ibrahim, R.A.
Vibro-Impact Dynamics
320 p. 2009 [978-3-642-00274-8]

Vol. 42: Hashiguchi, K.
Elastoplasticity Theory
400 p. 2009 [978-3-642-00272-4]

Vol. 41: Browand, F., Ross, J., McCallen, R. (Eds.)
Aerodynamics of Heavy Vehicles II: Trucks, Buses, and Trains
486 p. 2009 [978-3-540-85069-4]

Vol. 40: Pfeiffer, F.
Mechanical System Dynamics
578 p. 2008 [978-3-540-79435-6]

Vol. 39: Lucchesi, M., Padovani, C., Pasquinelli, G., Zani, N.
Masonry Constructions: Mechanical Models and Numerical Applications
176 p. 2008 [978-3-540-79110-2

Vol. 38: Marynowski, K.
Dynamics of the Axially Moving Orthotropic Web
140 p. 2008 [978-3-540-78988-8]

Vol. 37: Chaudhary, H.; Saha, S.K.
Dynamics and Balancing of Multibody Systems
200 p. 2008 [978-3-540-78178-3]

Vol. 36: Leine, R.I., van de Wouw, N.
Stability and Convergence of Mechanical Systems with Unilateral Constraints
250 p. 2008 [978-3-540-76974-3]

Vol. 35: Acary, V., Brogliato, B.
Numerical Methods for Nonsmooth Dynamical Systems: Applications in Mechanics and Electronics
545 p. 2008 [978-3-540-75391-9]

Vol. 34: Flores, P.; Ambrósio, J.; Pimenta Claro, J.C.; Lankarani Hamid M.
Kinematics and Dynamics of Multibody Systems with Imperfect Joints: Models and Case Studies
186 p. 2008 [978-3-540-74359-0]

Vol. 33: Nies ony, A.; Macha, E.
Spectral Method in Multiaxial Random Fatigue
146 p. 2007 [978-3-540-73822-0]

Vol. 32: Bardzokas, D.I.; Filshtinsky, M.L.; Filshtinsky, L.A. (Eds.)
Mathematical Methods in Electro-Magneto-Elasticity
530 p. 2007 [978-3-540-71030-1]

Vol. 31: Lehmann, L. (Ed.)
Wave Propagation in Infinite Domains
186 p. 2007 [978-3-540-71108-7]

Vol. 30: Stupkiewicz, S. (Ed.)
Micromechanics of Contact and Interphase Layers
206 p. 2006 [978-3-540-49716-5]

Vol. 29: Schanz, M.; Steinbach, O. (Eds.)
Boundary Element Analysis
571 p. 2006 [978-3-540-47465-4]

Vol. 28: Helmig, R.; Mielke, A.; Wohlmuth, B.I. (Eds.)
Multifield Problems in Solid and Fluid Mechanics
571 p. 2006 [978-3-540-34959-4]

Vol. 27: Wriggers P., Nackenhorst U. (Eds.)
Analysis and Simulation of Contact Problems
395 p. 2006 [978-3-540-31760-9]

Vol. 26: Nowacki, J.P.
Static and Dynamic Coupled Fields in Bodies with Piezoeffects or Polarization Gradient
209 p. 2006 [978-3-540-31668-8]

Vol. 25: Chen C.-N.
Discrete Element Analysis Methods of Generic Differential Quadratures
282 p. 2006 [978-3-540-28947-0]

Numerics of Unilateral Contacts and Friction

Modeling and Numerical Time Integration in Non-Smooth Dynamics

Christian Studer

Springer

Dr. Christian Studer
IMES-Center of Mechanics,
ETH Zurich,
8092 Zurich,
Switzerland
E-mail: christian.studer@alumni.ethz.ch

ISBN: 978-3-642-01099-6 e-ISBN: 978-3-642-01100-9

DOI 10.1007/978-3-642-01100-9

Lecture Notes in Applied and Computational Mechanics ISSN 1613-7736
e-ISSN 1860-0816

Library of Congress Control Number: Applied for

Typeset & Cover Design: Scientific Publishing Services Pvt. Ltd., Chennai, India.

Printed on acid-free paper

9 8 7 6 5 4 3 2 1 0

springer.com

Preface

Mechanics provides the link between mathematics and practical engineering applications. It is one of the oldest sciences, and many famous scientists have left and will leave their mark in this fascinating field of research. Perhaps one of the most prominent scientists in mechanics was Sir Isaac Newton, who with his "laws of motion" initiated the description of mechanical systems by differential equations. And still today, more than 300 years after Newton, this mathematical concept is more actual than ever.

The rising computer power and the development of numerical solvers for differential equations allowed engineers all over the world to predict the behavior of their physical systems fast and easy in an numerical way. And the trend to computational simulation methods is still further increasing, not only in mechanics, but practically in all branches of science. Numerical simulation will probably not solve the world's engineering problems, but it will help for a better understanding of the mechanisms of our models.

With the rise of computer power, demands in mechanical modeling obtained an additional focus. In "pre-computational mechanics" scientists aimed at short and intelligent formulations, for example ordinary (linear) differential equations expressed in minimal sets of coordinates. The "post-computational mechanics" also considers questions like how to provide simple and general modeling structures which are also suited for a computer. A machine, which can process large amounts of data, but is limited within its intelligence, i.e. right the opposite of a human being. Questions arise how such different "mechanisms" like computers and human beings can interact and complement each other. An example of this change in demands is the concept of differential algebraic equations, which are probably not the suited tool for "hand-evaluation" but are one of the favored ways how computers can set up general structures for constrained mechanical problems.

The rising computer power has also brought concepts into focus, for which excellent theoretical framework existed but which have been so far too data intensive for practical evaluation. One such field of increasing research activity is non-smooth mechanics, which deals with mechanical systems whose time evolution is not smooth anymore. Such modeling approaches help for the better understanding of systems with friction and impacts, which are omnipresent in our world, but of which still little is known from the modeling and numerical treatment point of view. Perhaps the most important contribution in non-smooth mechanics originates from Jean Jacques Moreau in the late 80's. He replaced the equations of motion and the constraint equations known from classical mechanics by measure differential equations in combination with inequalities. Doing so, Moreau was able not only to deal with classical smooth mechanics but also to account for non-smooth events like stick-slip transitions or impacts. In this sense, Moreau did not extend classical mechanics to some non-smooth special cases but set up a complete new formulation which incorporates classical mechanics. His excellent theoretical work was accompanied by a numerical integration scheme, which links the mathematical framework with practical applications.

In my years as research assistant at the Center of Mechanics at the ETH Zurich I obtained the chance to study this relatively new field of mechanics. I tried to bridge the gap between highly theoretical mathematical theories in non-smooth mechanics and practical application in software codes, I tried to identify the structure of non-smooth mechanics which allowed for simple and elegant implementation codes. During this time, I was accompanied by many people whom I would like to thank for their kind support of my work, which resulted finally in this book. Especially I want to thank my supervisor Prof. Dr.-Ing. Dr.-Ing. habil. Christoph Glocker for guiding my research and for the many interesting discussions about mechanics and the non-smoothness of our world. Prof. Glocker managed to create a fruitful uncomplicated atmosphere which allowed me to follow my own ideas and to develop freely within his support. He was also present in difficult times and encouraged me with his enthusiasm when speaking about mechanics. His broad knowledge in non-smooth mechanics and his strong quest for simple, consistent and elegant mathematical formulations have been a valuable inspiration for me. I also would like to thank Prof. Dr. habil. Bernard Brogliato for reviewing my scientific work. Prof. Brogliato is well known in the non-smooth dynamics community and heads the research team BIPOP at the INRIA in France. Further thanks go to all my colleagues at the center of mechanics, who made the institute an interesting, pleasant and open minded workplace. Special thanks go to Dr. ir. habil. Remco Leine who assisted me with much knowledge and who also reviewed my work, to Ueli Aeberhard who gave me mathematic support, and to Michael Möller with whom I had many discussions about numerics and programming. Finally I want to thank my family and my friends and especially Nadja Gerber for providing the motivating social background and mental support.

Zurich, 2008 Christian Studer

Contents

Chapter 1
Introduction

1.1 Aim and Scope

Reality is neither smooth nor non-smooth. Models of reality are smooth or non-smooth depending on the questions we ask. If a physical system has rapidly changing phases, then it can be advantageous to model the system in a non-smooth way. For mechanical systems the impact times are usually much smaller than the global motion which is of interest. This motivates the study of rigid multibody dynamics. The set-valued force laws which model the constitutive behaviour of unilateral contacts and of friction lead to non-smooth models. Usually, the positions are assumed to be absolutely continuous, while the velocities are allowed to undergo jumps and are taken to be of bounded variation. Jumps in the velocities can not be accomplished by finite but only by impulsive forces.

A point mass falling down to the ground is a simple example of a mechanical system with one unilateral contact. In a planar non-smooth modeling, the point mass's two degrees of freedom are reduced to one when the point mass touches the ground. If friction is considered additionally, then the degrees of freedom are reduced to zero in the case of sticking. Thus, there are different equations of motion in minimal coordinates for these different configurations. In case of an impact, an additional impact law must be applied.

This book focusses on numerical time integrators for the dynamics of non-smooth mechanical models. Such models consist of a set of rigid bodies which may interact with each other. Similar to classical smooth models, the state of a non-smooth mechanical model is described by a set of (minimal or non-minimal) generalized positions $\mathbf{q} = \mathbf{q}(t)$ and generalized velocities $\mathbf{u} = \mathbf{u}(t)$. Unlike smooth models, the time evolution of these states $\mathbf{q}$ and $\mathbf{u}$ is not smooth. While the generalized positions $\mathbf{q}$ are at least continuous, the generalized velocities $\mathbf{u} = \dot{\mathbf{q}}$ a.e.[1] may even jump at certain time instances. As a consequence, the accelerations might not be defined. Such non-smooth behaviour results from set-valued force laws which act

[1] The abbreviation a.e. stands for almost everywhere. It indicates that $\dot{\mathbf{q}} = \mathbf{u}$ does not hold at single time instances at which impacts occur.

C. Studer: Numerics of Unilateral Contacts and Friction, LNACM 47, pp. 1–8.
springerlink.com

between the rigid bodies. Similar to bilateral geometrical and kinematical constraints, set-valued force laws formulate restrictions on the states $\mathbf{q}$ and $\mathbf{u}$. Unlike bilateral constraints, these restrictions are not equations but inequalities. These inequalities define a (closed) set of admissible states $\mathbf{q}$ and $\mathbf{u}$, which has a normally an interior and a boundary. The non-smooth evolution of $\mathbf{q}$ and $\mathbf{u}$ is associated with sharpbends at time instances at which the states switch from the interior part of the admissible set to the boundary. Consider again the point mass which falls on the ground. The unilateral contact prevents penetration into the ground, i.e. it restricts the point mass's vertical position y by $y \geq 0$. Consider the situation right before the impact. At this time instance the position y is larger than zero. The velocity $u = \dot{y}$ a.e. grows linearly due to the gravitational force, and the displacements follow a parabolic path. At the moment of touching, the velocities must jump as else penetration would occur. As a consequence, also the time evolution of y becomes non-smooth. Such a velocity jump from the pre-impact velocity u^- to the post-impact velocity u^+ is called impact. Note that every post-impact velocity u^+ which avoids penetration into the ground is admissible. In order to choose a specific admissible post-impact velocity u^+, an impact law must be applied.

Note the following naming conventions in this book: A unilateral contact prevents penetration between bodies, where no friction is considered. A frictional unilateral contact prevents penetration between bodies, whereas friction is present in the tangent contact plane. The term set-valued force law includes the non-smooth representations of unilateral contacts, friction elements or other non-smooth interactions like pre-stressed springs. Note that a frictional unilateral contact is modeled within this book by two set-valued force laws, one for the unilateral and one for the frictional behaviour. If impulsive forces are taken into account, then we will speak about set-valued laws. A further convention applies to the term non-smooth mechanical system. This term is associated with the non-smooth model of a mechanical system and not the real system itself.

Non-smooth mechanical modeling can be found in various fields of technical application. Consider for example thousand balls which fall into a funnel. In this case, about half a million frictional unilateral contacts have to be evaluated in each time step in order to obtain the time evolution. Such granular media applications are of interest when investigating mixing processes. Other granular media models aim at predicting the behaviour of avalanches and rock fall. Apart from these granular systems, which incorporate many thousands of contacts, non-smooth modeling with a fewer number of contacts is of interest in virtually all parts of machine mechanics. Consider for example machines with linear joints which have limiting stops, or self-induced oscillations due to friction which may cause machine parts to squeal. The curve squealing of railroads or the brake squealing must be seen in this context. Further applications are found in the automotive and railroad industry, for which the contact between wheels and ground is essential. Non-smooth modeling might also be of interest when analyzing clockworks, especially the escapement. Also in robotics non-smooth models are applicable, consider for example walking robots or manipulators which interact which other bodies via unilateral contacts and/or friction. In control, the non-smooth modeling plays an essential role. The design of an advanced controller requests a good understanding of the occurring frictional and contact phenomena. In the past, non-smooth phenomena have been neglected if possible because no mathematical framework was available to describe such systems. If required, the non-smooth behaviour has been investigated in time intensive experimental work. In the recent years, the interest in non-smooth modeling has increased enormously. A mathematical framework has been developed which allows for an appropriate numeric evaluation of such systems. The aim is to handle the influence of unilateral contacts and/or friction on a mechanical system. By now the main interest is to avoid problems like squealing or undesired blocking. But non-smooth effects have much more to offer than problems which must be avoided. A better understanding of the influence of unilateral contacts and friction on mechanical systems allows for designing machines which intentionally use these effects for new manipulations. Note especially that human walking would not be possible without friction, as everyone can verify by trying to walk on ice. Projects which regard the non-smoothness as a chance and not as an impediment are for example the design of underactuated robots [98], walking machines, special mixing procedures, friction dampers or rattle conveyors. There exists also a nice variety of toys which use unilateral contacts and friction to produce astonishing movements, take for instance the woodpecker toy, the waddling duck, Euler's disk, the tippe-top, the rattleback or the super ball. In this sense, the non-smooth modeling of systems with unilateral contacts and friction should not only be seen as a method to avoid unwanted problems but also as tool to develop a new field of advanced mechanisms which intentionally use unilateral contacts and friction to fulfill their job.

This book aims at a better understanding and at an improvement of numerical methods for the simulation of non-smooth models. The focus is mainly on existing time-stepping integration methods, which are proven to be very robust in application. The book is based on several publications of the author, namely

- *A time-stepping method for non-smooth mechanical systems* [89], proceeding of the GAMM-conference 2005, which gives a short overview of Moreau's time-stepping scheme and its application.
- *Simulation of non-smooth mechanical systems with many unilateral constraints* [90], proceeding of the ENOC-conference 2005, which provides a more detailed discussion of Moreau's time-stepping scheme and which suggests step-size adjustment to locate switching points.
- *Representation of normal cone inclusion problems in dynamics via nonlinear equations* [91], journal publication which discusses the representation of non-smooth systems by projective equations, which are related to the augmented Lagrangian approach.
- *Solving normal cone inclusion problems in contact mechanics by iterative methods* [93], proceeding and selected journal publication of the ACMD-conference 2006, deals also with the representation of non-smooth mechanical systems by projective equations and their solution.
- *Augmented time-stepping by step size adjustment and extrapolation* [92], proceeding of the IDETC/CIE conference 2007 which gives a short overview on the

possibility to increase the accuracy of Moreau's time-stepping scheme by applying step-size adjustment and extrapolation.
- *Step size adjustment and extrapolation for time-stepping schemes in non-smooth dynamics* [94], journal publication which discusses in detail the application of step size adjustment and extrapolation to Moreau's time-stepping scheme.

1.2 Literature Survey

The numerical simulation of mechanical systems with unilateral contacts and friction covers a wide range of different topics. A first basic topic is the modeling of unilateral contacts and friction, which is essential for the simulation methods. Either one regularizes the associated force laws, i.e. the physical behaviour of unilateral contacts and friction is modeled by nonlinear stiff springs and dashpots, or one accepts the set-valued nature of the force laws and uses *set-valued* force laws to describe the physical behaviour. These set-valued force laws are represented by normal cone inclusions which are derived from a non-smooth potential. A next topic covers discretization schemes for the equations of motion in connection with the chosen force law model. Regularized force laws result in stiff ordinary differential equation (ODE) systems, which are solved straight forward with an ODE integrator for stiff systems. On the other hand, set-valued force laws lead to systems which are similar to differential algebraic equations (DAE), and which are integrated by event-driven or time-stepping approaches. A third topic is the handling of the inclusions which appear in both event-driven and time-stepping approaches. Either one writes these inclusions as complementarity problems or one treats them by an augmented Lagrangian approach. Finally, a short overview on applications, which have been calculated using time-stepping methods, is given. A review paper which gives a good overview on the whole topic and which lists over 200 references is given by Brogliato, ten Dam, Paoli, Génot and Abadie [22].

1.2.1 Modeling of Unilateral Contacts and Friction

Force laws for unilateral contacts and friction can either be regularized to become impressed forces, or they can be modeled as set-valued force laws. Consider for example the work of Stronge [87], in which springs are used to model impacts between rigid bodies. He proposes a compression and an expansion phase and uses different spring stiffnesses to obtain hysteresis. In [21], Stronge proposes an extended Hertz theory to obtain the corresponding spring stiffnesses. There exists a wide range of literature on this topic, a complete overview of which can not be given in this work. Regularized force laws have serious disadvantages in application, i.e. high frequency oscillations induced by the regularization may occur. The choice of the right parameters, i.e. spring stiffnesses etc. becomes essential.

Concerning the simulation process, regularized force laws lead to stiff ordinary differential systems, which must be solved iteratively. Nevertheless, regularized force laws are easy to implement in software packages, such as the multibody software packages Adams or Silux. This book deals solely with the non-smooth modeling of unilateral contacts and friction by set-valued force laws. Fundamental work has been done by Moreau [65], who formulated unilateral contact and friction laws as inclusion problems. The non-smooth modeling of unilateral contacts and friction is also discussed by various researchers, consider for example the works of Ballard, Brogliato, Frémond, Jourdain, Jean and Alart [13, 20, 35, 49, 50], Klarbring, Christensen, Pang and Stromberg [24,52] as well as Trinkle, Pang, Sudarsky and Lo [96]. This book is based on the contributions of Glocker and Pfeiffer [39, 41, 73], which discuss set-valued force laws as well as set-valued impact models. Two different impact models are proposed, the Newton impact model, which relates the pre- and post-impact velocities, and the Poisson impact model, which uses a compression and an expansion impulse. Further investigations on impact laws are made for example by Aeberhard and Payr [7] or Seifried [80]. Non-smooth mechanics is closely related to the topic of convex analysis, see Rockafellar [77]. A good summary on convex analysis is given by Glocker [39]. A comparison between the different contact models of Glocker and Stronge is provided by Payr [70].

1.2.2 *Time Discretization*

Standard literature on numerical integration is provided by Hairer and Wanner [45, 46], as well as in Strehmel [86]. Basic work concerning the discretization of differential algebraic equations with index three has been performed by Baumgarte, which introduced the so-called Baumgarte stabilization. A more advanced approach comes from Gear, Gupta and Leimkuhler, whose so-called GGL-method works with additional Lagrangian multipliers. A good overview on both methods is given in [46] or [86]. There exists a wide range of literature on the topic of numerical integration of differential algebraic equations. We exemplarily mention the work of Betsch [16, 17], who uses a discrete null space method to eliminate the constraints, the work of Botasso [19], who uses preconditioning to solve differential algebraic equations with index 3 straightforward, and the book of Eich-Soellner and Führer [30]. The Mexx integrator [59] for constrained mechanical systems comes from Lubich, Nowak, Pöhle and Engstler. This well known integrator uses extrapolation on a half explicit Euler method, and handles kinematic as well as geometric constraints on velocity level.

The two main classes of integrators for non-smooth mechanical systems are the event-driven and the time-stepping schemes. Event-driven schemes are discussed for example by Glocker and Pfeiffer in [39, 73]. These schemes integrate the equations of motion until a change in the state of a set valued force law is detected, i.e. a slip-stick transition or an impact. At these so-called switching points, the new states of the set-valued force laws are determined and the positions $\mathbf{q}$ and velocities $\mathbf{u}$

are re-initialized. Another type of integrator form the time-stepping methods, which are the main topic of this work. A first time-stepping method was introduced by Moreau [65]. Convergence proofs have been presented by Monteiro Marques [63] and by Dzonou [28]. Based on the time-stepping method of Moreau, various integration schemes have been developed. Jean [47] adapted the method to the so-called Θ-method, to which a similar method has been developed by Funk [36]. Funk also discusses the stability of his time-stepping method. Integration schemes which work on displacement level have been presented by Stiegelmeyr [85], Foerg et al. [33] and Pfeiffer et al. [72]. The authors also discuss stability and accuracy of their methods. Another method which models the set-valued force laws on displacement level has been developed by Paoli and Schatzmann in [68, 69]. They also give a proof for the convergence of their method. Acary and Brogliato discuss in [2] how nested complementarity problems can be used to incorporate geometric set-valued force laws into Moreau's time-stepping scheme. Further time-stepping methods are presented by Stewart and Trinkle [84] as well as Anitescu, Potra and Gavrea [12, 11, 74, 37]. Anitescu and Hart propose in [9] a method which uses drift stabilization. This drift stabilization must be seen in closer connection to the methods of Jean, Funk, Stiegelmeyr, Foerg and Pfeiffer. Ebrahimi and Eberhard propose in [29] to use Moreau's time-stepping scheme for flexible multibody systems. A separate overview on the different time-stepping methods is given in Chap. 5.

At last some references concerning extrapolation methods are given. These methods are used in this work to increase the integration order of Moreau's time-stepping schemes. An overview on the methods can be found in [45]. Deuflhard [26] and Lubich [58] apply the method on differential algebraic equations, see also [59].

1.2.3 Solvers for Inequality Problems

The non-smooth formulation of the contact problem results in normal cone inclusions, which can either be transformed into linear and nonlinear complementarity problems, or they can be solved by an augmented Lagrangian approach. Standard literature on linear complementarity problems is provided by Cottle and Pang [25], as well as Murty [67]. The authors describe various kind of complementarity problems, and provide solution methods, for example the Lemke's method. Fundamental work on how to formulate linear complementarity problems in non-smooth dynamics has been carried out by Glocker [39,40,41,42]. Funk proposed in [36] a modified Lemke's method which considers the mechanical nature of the problem. The augmented Lagrangian method was first proposed by Hesteness and Powell, see for example [15, 60, 75, 76]. The method was established in contact-mechanics by Alart and Curnier [8] and Simo and Laursen [82]. The method has proven to be very suitable, and it is used by a wide scope of researchers, for example Glocker [42], Leine [57] and Foerg [33]. A comparison between the complementarity and the augmented Lagrangian approach is done by Foerg, Pfeiffer and Ulbrich in [33,72]. Convergence strategies for the augmented Lagrangian iteration are done by Foerg [32] and by Studer [91]. Other approaches, i.e. Moreau's sweeping process [65], the successive Gauss-Seidel like solution of set-valued force laws [47,49] or the concept of exact regularization [57] are more or less identical to the augmented Lagrangian approach [91].

1.2.4 Application

Non-smooth modeling of mechanical systems is of increasing interest. Moreau uses time-stepping methods to simulate the behaviour of granular materials [66]. Lanier and Jean verify calculated granular applications by experiments [53]. Bohatier and Nouguier numerically investigate contact dynamics between a tool and granular material [18]. Impressive simulations of masonry structures, i.e. cathedrals, bridges or domes made of bricks, are done by Acary, Jean and Monerie [3,4,1,48]. Dumont and Paoli used their time-stepping method to model a vibrating beam between stops [27]. Abadie shows in [21] results from the simulation of a circuit breaker. Much work in non-smooth modeling of industrial applications has been done by Pfeiffer, Ulbrich, Foerg, Geier et al. We exemplarily mention the simulation of valve trains [34], push belt CVT's [38], roller coasters and drop towers [72]. Glocker, Leine and Le Saux use time-stepping methods for various non-smooth systems, i.e. the woodpecker toy [44], the tippe-top [56] or the rolling Euler disk [54]. Cataldi [23] has determined the frequency of the curve squealing for railroad tracks, while Transeth [95] has used time-stepping methods to simulate a robot snake. Widmer [99] uses both set-valued and regularized force laws to simulate a friction driven motor. Time-stepping methods have also successfully been used by Moeller and Glocker for the simulation of non-smooth electrical networks, see [62]. Slavic and Boltezar analyze a sliding pin which comes to contact with a rough rotating body [83]. Also Steward, Trinkle, Anitescu and Potra et al. provide some applications which have been calculated by their scheme [84,12,11,74,37]. A software code which incorporates the concepts of non-smooth dynamics is developed within the Siconos project [5,81]. Much work on the topic is also done in the computer graphics community, consider for example the work of Baraff [14] or Weinstein [97]. Non-smooth modeling becomes also essential in optimal control. We exemplarily mention the work of Yunt [100], who uses Moreau's time-stepping scheme in connection with optimization algorithms.

1.3 Outline

Chapter 2 gives an overview about the mathematics which is used in this book. Especially the introduction to convex analysis and the discussion of local and global representations of inclusion problems might be of interest for non-specialized readers.

A general non-smooth model for mechanical systems with unilateral contacts and friction is set up in Chap. 3, which starts with a short discussion of the equations of motion subjected to bilateral constraints expressed on different kinematic levels. Set-valued force laws are introduced, and the theory is extended to the non-smooth case, which includes also impacts. The chapter provides examples of set-valued force laws for selected non-smooth interactions, and shows different formal representations of a non-smooth mechanical system. Note that the dynamical behaviour of a non-smooth system can not be described by differential equations due to the non-smoothness of its states $\mathbf{q}$ and $\mathbf{u}$. Therefore, so-called equalities of measures are formulated, which provide a comprehensive analytic description of a non-smooth mechanical system.

Chapter 4 shows how inclusion problems can be solved. Such problems can not be circumvented when dealing with non-smooth systems. The solution process is discussed regarding inclusions which describe the impact-free motion on acceleration level, i.e. inclusions which describe the dynamical take-off or slip-stick transitions. Other inclusion problems, which occur when setting up impact laws or time-stepping schemes, have the same structure and can be approached in the same manner. The chapter gives a very brief overview on complementarity problems. Then, a general introduction to the augmented Lagrangian approach is given, which is afterwards applied to the non-smooth mechanical system. Doing so, it is possible to transform the inclusions into projective equations, which can be solved in an iterative way. Some iterative strategies are given, and other solution methods, which are associated with the augmented Lagrangian approach, are discussed.

Chapter 5 deals with different existing time-stepping schemes. First, the discretization of differential algebraic equations is investigated, whereas special attention is given to drift problems. Then the time evolution of non-smooth systems is analyzed more closely, and the main ideas of event-driven and time-stepping integration methods are outlined. The chapter focusses on the time-stepping methods, a selection of which is presented and categorized. A concluding discussion shows advantages and disadvantages of the different methods.

In Chap. 6, the accuracy of Moreau's time-stepping method is increased by using step size adjustment and extrapolation. The integration order of Moreau's time-stepping scheme is analyzed by observing smooth parts of the motion and switching incidents separately. An easy to manage and uniquely detection algorithm for switching points is presented, and a step size adjustment is proposed to resolve such time instances with a minimal step size. Within smooth parts of the motion, extrapolation methods are used to increase the integration order of Moreau's midpoint rule. Examples which use this augmented time-stepping algorithm are presented at the end of the chapter.

Finally, Chap. 7 gives a short overview on the C++ class structure which can be used to model non-smooth systems. In addition, some examples, which have been calculated using this C++ class structure, are presented.

Chapter 2
Mathematical Preliminaries

This chapter aims at giving a short introduction to some mathematical concepts which are used in this work. Vector and matrix norms are discussed in Sect. 2.1, some comments concerning derivatives are given in Sect. 2.2 and a short introduction to convex analysis is provided in Sect. 2.3. A proximal point and a distance vector function are introduced, which are associated with a special class of normal cone inclusion problems. A discussion on global and local representations of such problems is given in Sect. 2.4. Section 2.5 deals with differential algebraic equations.

2.1 Used Norms

In this work different vector and associated matrix norms are used [79]. The most popular norm is the Euclidian vector norm and its associated spectrum matrix norm. The definitions of these norms are

$$||\xi||_2 = \sqrt{\xi^\top \xi}\,, \quad ||\mathbf{A}||_2 = \sqrt{\rho(\mathbf{A}^\top \mathbf{A})}\,, \quad ||\mathbf{A}\xi||_2 \leq ||\mathbf{A}||_2 ||\xi||_2\,, \tag{2.1}$$

in which $\rho(\mathbf{A}^\top \mathbf{A})$ denotes the spectral radius of $\mathbf{A}^\top \mathbf{A}$, i.e. $\rho(\mathbf{A}^\top \mathbf{A}) = \max_i |\mu_i(\mathbf{A}^\top \mathbf{A})|$ with μ_i being the eigenvalues of $\mathbf{A}^\top \mathbf{A}$. Another norm is the maximum vector norm and its associated row-sum matrix norm, which is defined by

$$||\xi||_\infty = \max_h |\xi_h|\,, \quad ||\mathbf{A}||_\infty = \max_h \sum_{k=1}^{n} |a_{hk}|\,, \quad ||\mathbf{A}\xi||_\infty \leq ||\mathbf{A}||_\infty ||\xi||_\infty\,. \tag{2.2}$$

Furthermore, the norm

$$||\xi||_\mathrm{R} = \sqrt{\xi^\top \mathbf{R}\xi}\,, \quad ||\mathbf{A}||_\mathrm{R} = \sqrt{\rho(\mathbf{A}^\top \mathbf{R}\mathbf{A})}\,, \quad ||\mathbf{A}\xi||_\mathrm{R} \leq ||\mathbf{A}||_\mathrm{R} ||\xi||_\mathrm{R} \tag{2.3}$$

C. Studer: Numerics of Unilateral Contacts and Friction, LNACM 47, pp. 9–16.
springerlink.com

is used, in which $\mathbf{R}$ is *symmetric* and *positive definite*, i.e. $\xi^\top \mathbf{R}\xi > 0 \forall \xi \neq 0$. Note that arbitrary matrix norms $||\mathbf{A}||$ provide an upper bound for $\rho(\mathbf{A})$, i.e. $||\mathbf{A}|| \geq \rho(\mathbf{A})$. Furthermore, there exists a matrix norm $||\cdot||_\star$ with $||\mathbf{A}||_\star = \rho(\mathbf{A})$. This matrix norm is referred as minimal matrix norm since $||\mathbf{A}|| \geq ||\mathbf{A}||_\star = \rho(\mathbf{A})$. In case of $\mathbf{A} = \mathbf{A}_{\text{sym}}$ being symmetric, the spectrum norm $||\cdot||_2$ provides this minimal matrix norm, i.e. $||\mathbf{A}_{\text{sym}}||_2 = ||\mathbf{A}_{\text{sym}}||_\star = \rho(\mathbf{A})$.

2.2 Derivatives

The following conventions concerning derivatives are used in this work:

$$\frac{\partial f}{\partial \mathbf{x}} = \left(\frac{\partial f}{\partial x_1} \cdots \frac{\partial f}{\partial x_m}\right), \qquad \frac{\partial \mathbf{f}}{\partial \mathbf{x}} = \begin{pmatrix} \frac{\partial f_1}{\partial \mathbf{x}} \\ \vdots \\ \frac{\partial f_n}{\partial \mathbf{x}} \end{pmatrix} = \begin{pmatrix} \frac{\partial f_1}{\partial x_1} & \cdots & \frac{\partial f_1}{\partial x_m} \\ \vdots & \ddots & \vdots \\ \frac{\partial f_n}{\partial x_1} & \cdots & \frac{\partial f_n}{\partial x_m} \end{pmatrix}. \tag{2.4}$$

The chain rule reads as

$$\frac{\partial \mathbf{f}(\mathbf{k}(\mathbf{g}(\mathbf{x})))}{\partial \mathbf{x}} = \frac{\partial \mathbf{f}(\mathbf{k})}{\partial \mathbf{k}} \frac{\partial \mathbf{k}(\mathbf{g})}{\partial \mathbf{g}} \frac{\partial \mathbf{g}(\mathbf{x})}{\partial \mathbf{x}}, \tag{2.5}$$

for example

$$\frac{\partial}{\partial \mathbf{x}} \frac{1}{2} ||\mathbf{k}(\mathbf{g}(\mathbf{x}))||_{\mathrm{R}}^2 = \mathbf{k}(\mathbf{g}(\mathbf{x}))^\top \mathbf{R} \frac{\partial \mathbf{k}(\mathbf{g})}{\partial \mathbf{g}} \frac{\partial \mathbf{g}(\mathbf{x})}{\partial \mathbf{x}}. \tag{2.6}$$

2.3 Convex Analysis

This section provides a brief introduction to convex analysis, of which the knowledge is essential to understand the contact modeling. The reader is referred to [39, 77].

2.3.1 *Convex Sets, Indicator / Support Functions, Normal Cones*

A convex set $\mathscr{C}$ is a set of points $\mathbf{x} \in \mathbb{R}^n$ whose connection lies also in the set $\mathscr{C}$, i.e. it holds for all $\mathbf{x}_1 \in \mathscr{C}$ and $\mathbf{x}_2 \in \mathscr{C}$ that $\mathbf{x}_3 = (1-\alpha)\mathbf{x}_1 + \alpha \mathbf{x}_2 \in \mathscr{C}, 0 \leq \alpha \leq 1$. The indicator function $\Psi_{\mathscr{C}}(\mathbf{x})$ of a convex set $\mathscr{C}$ takes the value zero if $\mathbf{x} \in \mathscr{C}$ and infinity otherwise,

$$\Psi_{\mathscr{C}}(\mathbf{x}) = \begin{cases} 0 \ \text{if } \mathbf{x} \in \mathscr{C} \\ \infty \ \text{if } \mathbf{x} \notin \mathscr{C}\,. \end{cases} \tag{2.7}$$

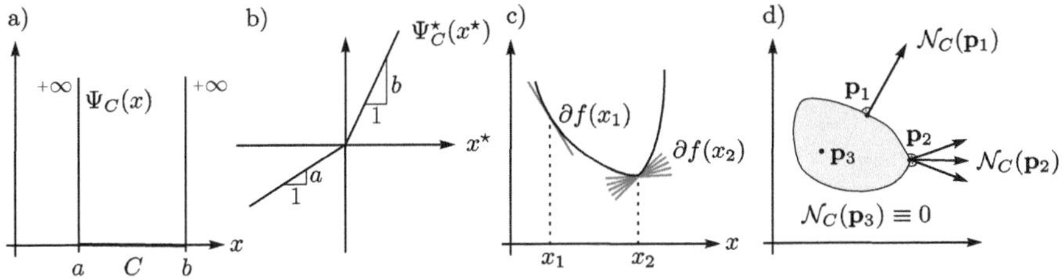

Fig. 2.1 Indicator (**a**) and support function (**b**), subdifferential (**c**) and normal cone (**d**) [39,93]

The indicator function of a convex set $\mathscr{C} = [a,b]$ is depicted in Fig. 2.1a. The conjugate to this indicator function is the support function $\Psi^\star_{\mathscr{C}}(\mathbf{x}^\star)$,

$$\Psi^\star_{\mathscr{C}}(\mathbf{x}^\star) = \sup_{\mathbf{x}}\{\mathbf{x}^\top \mathbf{x}^\star \,|\, \mathbf{x} \in \mathscr{C}\}, \tag{2.8}$$

which is shown in Fig. 2.1b. The subdifferential $\partial f(\mathbf{x})$ of a convex function $f(\mathbf{x})$ is a set containing the gradients of all supporting hyperplanes of $f(\mathbf{x})$,

$$\partial f(\mathbf{x}) = \{\mathbf{y} \,|\, f(\mathbf{x}^\star) \geq f(\mathbf{x}) + \mathbf{y}^\top(\mathbf{x}^\star - \mathbf{x}); \forall \mathbf{x}^\star\}, \tag{2.9}$$

see also Fig. 2.1c. Note that the subdifferential is defined as a column vector, opposed to the derivative $\frac{\partial f}{\partial \mathbf{x}}$ which comes as a row vector. The subdifferential of the indicator function $\partial \Psi_C(\mathbf{x})$ is identical to the normal cone $\mathscr{N}_{\mathscr{C}}(\mathbf{x})$, which is the set of all vectors making an obtuse angle with all other vectors emanating from $\mathbf{x} \in \mathscr{C}$ to any $\hat{\mathbf{x}} \in \mathscr{C}$,

$$\mathscr{N}_{\mathscr{C}}(\mathbf{x}) = \{\mathbf{y} \,|\, \mathbf{y}^\top(\mathbf{x}^\star - \mathbf{x}) \leq 0; \forall \mathbf{x}^\star \in \mathscr{C}; \mathbf{x} \in \mathscr{C}\}. \tag{2.10}$$

The equivalence $\mathscr{N}_{\mathscr{C}}(\mathbf{x}) \equiv \partial \Psi_{\mathscr{C}}(\mathbf{x})$ can be verified by comparing the definitions (2.9) and (2.10), where $\mathbf{f}(\mathbf{x}) = \Psi_{\mathscr{C}}(\mathbf{x}) = 0$ if $\mathbf{x} \in \mathscr{C}$. Figure 2.1d depicts the normal cone to a set $\mathscr{C}$ at different points $\mathbf{p}_i$. Furthermore, the subdifferential of the indicator function is the inverse of the subdifferential of the support function,

$$\mathbf{x}^\star \in \partial \Psi_C(\mathbf{x}) \quad \Leftrightarrow \quad \mathbf{x} \in \partial \Psi^\star_C(\mathbf{x}^\star). \tag{2.11}$$

2.3.2 *Proximal Point and Vector Distance Functions*

In the following, a proximal point and a vector distance function are defined. The proximal point function $\mathrm{prox}^{\mathrm{R}}_{\mathscr{C}}(\cdot)$ performs a projection onto the set $\mathscr{C}$, i.e. the point $\mathbf{x} = \mathrm{prox}^{\mathrm{R}}_{\mathscr{C}}(\xi)$ is the nearest point to ξ in the set $\mathscr{C}$ with respect to $||\cdot||_{\mathrm{R}}$,

$$\mathbf{x} = \mathrm{prox}^{\mathrm{R}}_{\mathscr{C}}(\xi) = \underset{\mathbf{x}^\star \in \mathscr{C}}{\mathrm{argmin}}\, \frac{1}{2}||\xi - \mathbf{x}^\star||^2_{\mathrm{R}}. \tag{2.12}$$

The distance vector function $\mathrm{vdst}_{\mathscr{C}}^{\mathrm{R}}(\cdot)$ returns the shortest vector emanating from the set $\mathscr{C}$ to an arbitrary point ξ with respect to $||\cdot||_{\mathrm{R}}$,

$$\mathrm{vdst}_{\mathscr{C}}^{\mathrm{R}}(\xi) = \xi - \mathrm{prox}_{\mathscr{C}}^{\mathrm{R}}(\xi). \tag{2.13}$$

Note that the term $\frac{1}{2}||\mathrm{vdst}_{\mathscr{C}}^{\mathrm{R}}(\xi)||_{\mathrm{R}}^2$ is continuously differentiable [55][1]

$$\left(\frac{\partial}{\partial \xi}\frac{1}{2}||\mathrm{vdst}_{\mathscr{C}}^{\mathrm{R}}(\xi)||_{\mathrm{R}}^2\right)^{\top} = \left(\frac{\partial\, \mathrm{vdst}_{\mathscr{C}}^{\mathrm{R}}(\xi)}{\partial \xi}\right)^{\top}\mathbf{R}\,\mathrm{vdst}_{\mathscr{C}}^{\mathrm{R}}(\xi) = \mathbf{R}\,\mathrm{vdst}_{\mathscr{C}}^{\mathrm{R}}(\xi). \tag{2.14}$$

In case of $\mathbf{R}$ being the identity matrix $\mathbf{I}$, the upper index ^{R} is omitted and the proximal point and the distance vector function are written as $\mathrm{prox}_{\mathscr{C}}$ and $\mathrm{vdst}_{\mathscr{C}}$. Note that scaling the identity matrix $\mathbf{I}$ by a positive scalar α has no influence on the proximal point and the distance vector function, i.e. $\mathrm{prox}_{\mathscr{C}}^{\alpha\mathrm{I}} = \mathrm{prox}_{\mathscr{C}}^{\mathrm{I}} = \mathrm{prox}_{\mathscr{C}}$.

The $\mathrm{prox}_{\mathscr{C}}^{\mathrm{R}}$ function is closely related to the normal cone $\mathscr{N}_{\mathscr{C}}$ (2.10). The restriction $\mathbf{x}^{\star} \in \mathscr{C}$ in (2.12) can be omitted by using the indicator function $\Psi_{\mathscr{C}}(\mathbf{x}^{\star})$, which yields the definition of the $\mathrm{prox}_{\mathscr{C}}^{\mathrm{R}}$ function to become

$$\mathbf{x} = \mathrm{prox}_{\mathscr{C}}^{\mathrm{R}}(\xi) = \underset{\mathrm{x}^{\star}}{\mathrm{argmin}}\, \frac{1}{2}||\xi - \mathbf{x}^{\star}||_{\mathbf{R}}^2 + \Psi_{\mathscr{C}}(\mathbf{x}^{\star}) =: \underset{\mathrm{x}^{\star}}{\mathrm{argmin}}\, f(\mathbf{x}^{\star}). \tag{2.15}$$

The argument $\mathbf{x}$ which minimizes $f(\mathbf{x}^{\star})$ follows by the subdifferential $\partial f(\mathbf{x}^{\star})$,

$$\partial f(\mathbf{x}) \equiv -\mathbf{R}(\xi - \mathbf{x}) + \partial\Psi_{\mathscr{C}}(\mathbf{x}) \equiv -\mathbf{R}(\xi - \mathbf{x}) + \mathscr{N}_{\mathscr{C}}(\mathbf{x}) \ni 0, \tag{2.16}$$

which yields the relation

$$\mathbf{x} = \mathrm{prox}_{\mathscr{C}}^{\mathrm{R}}(\xi) \quad \Leftrightarrow \quad \mathbf{x} + \mathbf{R}^{-1}\mathscr{N}_{\mathscr{C}}(\mathbf{x}) \ni \xi, \tag{2.17}$$

Replacing $\mathbf{R}$ by a scaled identity matrix $\frac{1}{r}\mathbf{I}$ with arbitrary $r > 0$ yields

$$\mathbf{x} = \mathrm{prox}_{\mathscr{C}}(\xi) \quad \Leftrightarrow \quad \mathbf{x} + r\mathscr{N}_{\mathscr{C}}(\mathbf{x}) \equiv \mathbf{x} + \mathscr{N}_{\mathscr{C}}(\mathbf{x}) \ni \xi. \tag{2.18}$$

In some cases it might be useful to take transformed convex sets $\mathscr{C}$, i.e. $\mathscr{C} \to \mathbf{A}^{-1}\mathscr{C}$, where $\mathbf{A}$ is positive definite. The following two equivalent relations might be of interest:

$$\mathrm{prox}_{\mathscr{C}}^{\mathrm{R}}(\xi) = \mathbf{A}\,\mathrm{prox}_{\mathrm{A}^{-1}\mathscr{C}}^{\mathrm{A}^{\top}\mathrm{RA}}(\mathbf{A}^{-1}\xi), \tag{2.19}$$

$$\mathscr{N}_{\mathscr{C}}(\mathbf{Ax}) = (\mathbf{A}^{\top})^{-1}\mathscr{N}_{\mathrm{A}^{-1}\mathscr{C}}(\mathbf{x}). \tag{2.20}$$

The two relations can be verified by consulting the definitions of the prox_C function (2.12) and of the normal cone (2.10). Choosing $\mathbf{R} = \mathbf{I}$ in (2.19) yields a connection between skewed and Euclidian projections,

[1] [55] proofs relation (2.14) only for $\mathbf{R} = \mathbf{I}$. Nevertheless, the proof can be extended to arbitrary symmetric and positive definite matrices $\mathbf{R}$ using (2.21).

$$\mathrm{prox}_{\mathscr{C}_1}(\xi_1) = \mathbf{A}\,\mathrm{prox}^{\mathbf{A}^\top\mathbf{A}}_{\mathbf{A}^{-1}\mathscr{C}_1}(\mathbf{A}^{-1}\xi_1) \quad \text{or} \quad \mathbf{A}^{-1}\mathrm{prox}_{\mathbf{A}\mathscr{C}_2}(\mathbf{A}\xi_2) = \mathrm{prox}^{\mathbf{A}^\top\mathbf{A}}_{\mathscr{C}_2}(\xi_2) \tag{2.21}$$

2.3.3 *Proximal Point Functions for Various Convex Sets*

Table 2.1 lists various convex sets with their associated proximal point functions. These sets will later be used to characterize the different set-valued force laws. Note

Table 2.1 Various convex sets with associated proximal point functions. The 4-order polynomial for the projection to the sets $\mathscr{S}_a$ and $\mathscr{S}_{cc}$ can be found in appendix B

Convex set	proximal point function
$\mathbb{R}_0^+$	$\mathrm{prox}_{\mathbb{R}_0^+}(\xi) = \begin{cases} \xi & \text{if } \xi \geq 0 \\ 0 & \text{if } \xi < 0 \end{cases}$
$\mathscr{S}_p = [-a, a]$	$\mathrm{prox}_{\mathscr{S}_p}(\xi) = \begin{cases} a & \text{if } \xi > a \\ \xi & \text{if } a \geq \xi \geq -a \\ -a & \text{if } \xi < -a \end{cases}$
$\mathscr{S}_s = \{\xi \in \mathbb{R}^2 \mid \|\|\xi\|\|_2 \leq a\}$	$\mathrm{prox}_{\mathscr{S}_s}(\xi) = \begin{cases} \xi & \text{if } \|\|\xi\|\|_2 \leq a \\ a\dfrac{\xi}{\|\|\xi\|\|_2} & \text{if } \|\|\xi\|\|_2 > a \end{cases}$
$\mathscr{S}_a = \{(\xi_1\ \xi_2)^\top \in \mathbb{R}^2 \mid \left(\frac{\xi_1}{c}\right)^2 + \left(\frac{\xi_2}{d}\right)^2 \leq 1\}$	requests solution of 4-order polynomial
$\mathscr{S}_{cc} = \{(\xi_1\ \xi_2\ \xi_3)^\top \in \mathbb{R}^3 \mid \left(\frac{\xi_1}{c}\right)^2 + \left(\frac{\xi_2}{c}\right)^2 + \left(\frac{\xi_3}{d}\right)^2 \leq 1\}$	requests solution of 4-order polynomial

that the convex sets $\mathscr{S}_a$ and $\mathscr{S}_{cc}$ have an ellipse and an ellipsoidal shape, respectively. The associated proximal point functions $\mathrm{prox}_{\mathscr{S}_a}$ and $\mathrm{prox}_{\mathscr{S}_{cc}}$ require therefore the solution of a forth order polynomial. The solution procedure is only sketched for the $\mathrm{prox}_{\mathscr{S}_a}$ function. It is easy to determine whether $\xi = (\xi_1\ \xi_2)^\top$ is within the set $\mathscr{S}_a$ or not. If $\xi \notin \mathscr{S}_a$, then a projection to the ellipse contour $e(\cdot) = 0$ has to be performed. It holds that

$$\mathbf{x} = \mathrm{prox}_{\mathscr{S}_a}(\xi) \quad \Rightarrow \quad \mathbf{x} + \alpha \nabla e(\mathbf{x}) = \xi, \quad e(\mathbf{x}) = 0, \tag{2.22}$$

where the gradient $\nabla e(\cdot)$ of the ellipse contour $e(\mathbf{x}) = 0$ is assumed to point in the outward direction. The condition $\mathbf{x} + \alpha \nabla e(\mathbf{x}) = \xi$ can be used to eliminate $\mathbf{x}$ from the condition $e(\mathbf{x}) = 0$, which yields a forth order polynomial for α. Solving this polynomial yields a positive, a negative and two imaginary solutions. The proximal point is calculated using the positive solution. The projection to an ellipsoid contour can be done in a similar way. Some intermediate calculation steps can be found in appendix B.

Note that until now only the Euclidian norm has been used for the projections. Considering projections to the sets $\mathscr{S}_a$ and $\mathscr{S}_{cc}$ with respect to arbitrary norms offers interesting possibilities. According to (2.21), such projections can also be viewed as Euclidian projections to transformed sets. Consider for example $\mathbf{x} = \text{prox}^{\text{R}}_{\mathscr{S}_a}(\xi)$, where $\mathbf{R}$ is free of choice within the set of all positive definit symmetric matrices, for example $\mathbf{R} = \mathbf{A}^\top\mathbf{A}$. Choosing $\mathbf{A}$ as a diagonal matrix for which $\mathbf{A}\mathscr{S}_a$ becomes a circular set, i.e. $A_{11} = 1$ and $A_{22} = \frac{c}{d}$, allows for replacing the skewed prox function $\mathbf{x} = \text{prox}^{\text{R}}_{\mathscr{S}_a}(\xi)$ to the ellipse set $\mathscr{S}_a$ by an equivalent Euclidian prox function $\mathbf{x} = \mathbf{A}^{-1}\text{prox}_{\mathbf{A}\mathscr{S}_a}(\mathbf{A}\,\xi)$ to a circular set.

2.4 Global and Local Representations

Let $\mathbf{x} \in \mathbb{R}^m$ and $\mathbf{y} \in \mathbb{R}^m$ be global vectors which are composed of n local vectors $\mathbf{x}_i \in \mathbb{R}^{m_i}$ and $\mathbf{y}_i \in \mathbb{R}^{m_i}$,

$$\mathbf{x} = (\mathbf{x}_1^\top \ldots \mathbf{x}_n^\top)^\top, \quad \mathbf{y} = (\mathbf{y}_1^\top \ldots \mathbf{y}_n^\top)^\top. \tag{2.23}$$

The scalars m and m_i denote the dimension of the global and of the i-th local problem, respectively. The scalar n stands for the number of local problems, i.e.

$$\sum_{i=1}^{n} m_i = m. \tag{2.24}$$

Let $\mathscr{C} \subset \mathbb{R}^m$ be a global convex set built up from the cartesian product of n local convex sets $\mathscr{C}_i \subset \mathbb{R}^{m_i}$,

$$\mathscr{C} = \mathscr{C}_1 \times \mathscr{C}_2 \times \ldots \times \mathscr{C}_n. \tag{2.25}$$

By doing so it is possible to write n local normal cone inclusions of the form

$$\mathbf{x}_i \in \mathscr{N}_{\mathscr{C}_i}(\mathbf{y}_i), \quad i = 1 \ldots n \tag{2.26}$$

as one global normal cone inclusion

$$\mathbf{x} \in \mathscr{N}_{\mathscr{C}}(\mathbf{y}). \tag{2.27}$$

In the same manner, n projections to local convex sets $\mathscr{C}_i$

$$\mathbf{x}_i = \text{prox}^{\text{R}_i}_{\mathscr{C}_i}(\xi_i) = \underset{\mathbf{x}_i^\star \in \mathscr{C}_i}{\text{argmin}}\, ||\xi_i - \mathbf{x}_i^\star||^2_{\mathbf{R}_i} = \underset{\mathbf{x}_i^\star \in \mathscr{C}_i}{\text{argmin}}\, (\xi_i - \mathbf{x}_i^\star)^\top \mathbf{R}_i (\xi_i - \mathbf{x}_i^\star), \quad i = 1 \ldots n \tag{2.28}$$

Table 2.2 Local representation versus the associated global representation of a normal cone inclusion problem

n local problems	global problem
$\mathbf{x}_i + \mathbf{R}_i^{-1}\mathscr{N}_{\mathscr{C}_i}(\xi_i) \ni \xi_i \ \Leftrightarrow\ \mathbf{x}_i = \mathrm{prox}_{\mathscr{C}_i}^{\mathrm{R}_i}(\xi_i)$	$\mathbf{x} + \mathbf{R}^{-1}\mathscr{N}_{\mathscr{C}}(\xi) \ni \xi \ \Leftrightarrow\ \mathbf{x} = \mathrm{prox}_{\mathscr{C}}^{\mathrm{R}}(\xi)$
$\mathbf{x}_i \in \mathbb{R}^{m_i}$	$\mathbf{x} = (\mathbf{x}_1 \ldots \mathbf{x}_n)^\top \in \mathbb{R}^m$
$\xi_i \in \mathbb{R}^{m_i}$	$\xi = (\xi_1 \ldots \xi_n)^\top \in \mathbb{R}^m$
$\mathscr{C}_i \subset \mathbb{R}^{m_i}$	$\mathscr{C} = \mathscr{C}_1 \times \ldots \times \mathscr{C}_n \subset \mathbb{R}^m$
$\mathbf{R}_i \in \mathbb{R}^{m_i \times m_i}$	$\mathbf{R} = \mathbf{R}_1 \oplus \ldots \oplus \mathbf{R}_n \in \mathbb{R}^{m \times m}$

are equivalent to a single projection to the global set $\mathscr{C}$,

$$\mathbf{x} = \mathrm{prox}_{\mathscr{C}}^{\mathrm{R}}(\xi) = \underset{\mathbf{x}^\star \in \mathscr{C}}{\mathrm{argmin}}\, ||\xi - \mathbf{x}^\star||_{\mathbf{R}}^2 = \underset{\mathbf{x}^\star \in \mathscr{C}}{\mathrm{argmin}} \sum_{i=1}^{n} (\xi_i - \mathbf{x}_i^\star)^\top \mathbf{R}_i (\xi_i - \mathbf{x}_i^\star), \quad (2.29)$$

if the matrix $\mathbf{R} \in \mathbb{R}^{m \times m}$ is a block matrix

$$\mathbf{R} = \mathbf{R}_1 \oplus \mathbf{R}_2 \oplus \ldots \oplus \mathbf{R}_n, \quad \mathbf{R}_i \in \mathbb{R}^{m_i \times m_i}. \quad (2.30)$$

The operator $\oplus$ stands for the direct sum between matrices, i.e.

$$\mathbf{A} \oplus \mathbf{B} = \begin{pmatrix} \mathbf{A} & 0 \\ 0 & \mathbf{B} \end{pmatrix}. \quad (2.31)$$

Note that (2.28) and (2.29) are only equivalent because the choice of an $x_i \in \mathscr{C}_i$ does not influence the choice $x_j \in \mathscr{C}_j$ due to (2.25). Table 2.2 relates the n local problems to the associated global problem.

2.5 Differential Algebraic Equations

A system which consists of differential and algebraic equations is called differential algebraic equation system (DAE). A characteristic property of differential algebraic equations is the differential index and the perturbation index. The differential index stands for the number of derivations which are are necessary to obtain an ordinary differential system. The perturbation index indicates the sensitivity of the solution of the DAE with respect to initial perturbations. In this book, only semi-explizite differential algebraic systems are used,

$$\mathbf{g}(t, \mathbf{x}, \lambda) = 0, \quad (2.32)$$

$$\dot{\mathbf{x}} - \mathbf{f}(t, \mathbf{x}, \lambda) = 0, \quad (2.33)$$

for which differential and perturbation index are the same. Note the algebraic equation (2.32) and the differential equation (2.33). If the algebraic constraint (2.32) is of the form

$$\mathbf{g}(t,\mathbf{x},\lambda)=0, \quad \frac{\partial \mathbf{g}}{\partial \lambda} \text{ is regular,} \tag{2.34}$$

then the differential index is one. This can be verified by differentiating the algebraic constraint (2.32) with respect to time t, which yields

$$\frac{\partial \mathbf{g}}{\partial t}+\frac{\partial \mathbf{g}}{\partial \mathbf{x}}\underbrace{\mathbf{f}(t,\mathbf{x},\lambda)}_{=\dot{\mathbf{x}}}+\frac{\partial \mathbf{g}}{\partial \lambda}\dot{\lambda}=0 \quad \Rightarrow \quad \dot{\lambda}=-\left[\frac{\partial \mathbf{g}}{\partial \lambda}\right]^{-1}\left(\frac{\partial \mathbf{g}}{\partial t}+\frac{\partial \mathbf{g}}{\partial \mathbf{x}}f(t,\mathbf{x},\lambda)\right), \tag{2.35}$$

i.e. one derivation of the algebraic constraint with respect to time t yields an ordinary differential equation system for $\mathbf{x}(t)$ and $\lambda(t)$. Consider the algebraic constraint

$$\mathbf{g}(t,\mathbf{x})=0, \tag{2.36}$$

for which $\frac{\partial \mathbf{g}}{\partial \lambda}=0$ is non-regular. If two derivations of (2.36) with respect to time t yield an explicit function $\dot{\lambda}=\mathbf{k}(t,\mathbf{x},\lambda)$, then the index of the DAE is two. An index 3 DAE requests three times differentiating the algebraic constraint (2.36). More detailed information on DAE's is available in [46, 86].

Chapter 3
Non-Smooth Mechanics

This chapter deals with the non-smooth modeling of mechanical systems. In Sect. 3.1 it is reviewed how a mechanical system subjected to bilateral constraints can be formulated as a differential algebraic equation (DAE). The theory is extended in Sect. 3.2 to non-impulsive non-smooth motion, and it is discussed how unilateral contacts, friction and other non-smooth interactions can be modeled by set-valued force laws. In Sect. 3.3 impacts and other impulsive interactions are added to the model by stating impact equations and Newton's extended impact law. Both impulsive and non-impulsive motion can be gathered together in so-called equalities of measures, which are discussed in Sect. 3.4. This chapter deals with rigid multibody systems. However, the resulting equations are much more general and can also be used for the non-rigid case.

3.1 Equations of Motion in the Smooth Case

Applying the principle of virtual work yields the equations of motion for the case without constraints

$$\mathbf{M}\dot{\mathbf{u}} = \mathbf{h}, \tag{3.1}$$

$$\dot{\mathbf{q}} = \mathbf{u}, \tag{3.2}$$

with $\mathbf{M} = \mathbf{M}(\mathbf{q},t)$ being the positive definite and symmetric mass matrix and $\mathbf{h} = \mathbf{h}(\mathbf{q},\mathbf{u},t)$ the vector of all external and gyroscopic forces acting on the system. Forces originating from springs or dashpots are also included in this vector $\mathbf{h}$. The vector $\mathbf{q} = \mathbf{q}(t)$ denotes the generalized displacements, the vector $\mathbf{u} = \mathbf{u}(t)$ addresses the generalized velocities. Next, constraints are considered to act on the system. Constraints are restrictions on the displacements $\mathbf{q}$ and/or the velocities $\mathbf{u}$ in equality form. If these restrictions are solely stated in terms of the displacements $\mathbf{q}$, then one speaks of a formulation on displacement level. If also velocities $\mathbf{u}$ are used, the formulation is assigned to be on velocity level. Finally, formulations on acceleration level relate positions $\mathbf{q}$, velocities $\mathbf{u}$ and accelerations $\dot{\mathbf{u}}$ to each other.

C. Studer: Numerics of Unilateral Contacts and Friction, LNACM 47, pp. 17–30.
springerlink.com

A *geometric* constraint i can always be described on displacement level, i.e. a joint between two rigid bodies. Such a constraint is characterized by a constraint equation $\kappa_i(\mathbf{q},t) = 0$. In order to relate the topic to unilateral contacts, the gap function $\mathbf{g}_i = \mathbf{g}_i(\mathbf{q},t)$ is chosen as constraint function , i.e. the geometric constraint should be described by

$$\mathbf{g}_i(\mathbf{q},t) = 0. \tag{3.3}$$

The gap function is defined such that its total time derivative yields the relative constraint velocity $\dot{\mathbf{g}}_i = \gamma_i = \mathbf{W}_i^\top \mathbf{u} + \zeta_i$ a.e., in which $\mathbf{W}_i = \mathbf{W}_i(\mathbf{q},t) = (\partial \mathbf{g}_i / \partial \mathbf{q})^\top$ and $\zeta_i = \zeta_i(\mathbf{q},t) = \partial \mathbf{g}_i / \partial t$. The total time derivative of the relative constraint velocity γ_i yields the relative constraint accelerations $\ddot{\mathbf{g}}_i = \dot{\gamma}_i = \mathbf{W}^\top \dot{\mathbf{u}} + \hat{\zeta}_i$ a.e., in which $\hat{\zeta}_i = \hat{\zeta}_i(\mathbf{q},\mathbf{u},t)$. Note that a geometric constraint can also be described on velocity level if the initial conditions are fulfilled on displacement level,

$$\mathbf{g}_i(\mathbf{q},t) = 0 \quad \Leftrightarrow \quad \dot{\mathbf{g}}_i = \gamma_i = \mathbf{W}_i^\top \mathbf{u} + \zeta_i = 0, \quad \mathbf{g}_i(\mathbf{q}_0,t_0) = 0. \tag{3.4}$$

In order to satisfy the constraint we place ourselves on a point of the manifold $\mathbf{g}_i = 0$ by requesting $\mathbf{g}_i(\mathbf{q}_0,t_0) = 0$. Furthermore, we prevent a drift from the manifold by demanding $\gamma_i = 0$. Finally, a geometric constraint can be described on acceleration level,

$$\mathbf{g}_i(\mathbf{q},t) = 0 \quad \Leftrightarrow \quad \ddot{\mathbf{g}}_i = \dot{\gamma}_i = \mathbf{W}_i^\top \dot{\mathbf{u}} + \hat{\zeta}_i = 0, \quad \gamma_i(\mathbf{q}_0,\mathbf{u}_0,t_0) = 0, \quad \mathbf{g}_i(\mathbf{q}_0,t_0) = 0. \tag{3.5}$$

A *kinematic* constraint naturally occurs on velocity level. Such a constraint is modeled by

$$\text{velocity level}: \quad \gamma_i = \mathbf{W}_i^\top \mathbf{u} + \zeta_i = 0, \tag{3.6}$$

$$\text{acceleration level}: \quad \dot{\gamma}_i = \mathbf{W}_i^\top \dot{\mathbf{u}} + \hat{\zeta}_i = 0, \; \gamma_i(\mathbf{q}_0,\mathbf{u}_0,t_0) = 0, \tag{3.7}$$

where $\mathbf{W}_i = \mathbf{W}_i(\mathbf{q},t)$, $\zeta_i = \zeta_i(\mathbf{q},t)$ and $\hat{\zeta}_i = \hat{\zeta}_i(\mathbf{q},\mathbf{u},t)$.

Consider a mechanical system subjected to n constraints, which are linked with the equations of motion by a Lagrange-I formulation in the form

$$\mathbf{M}\dot{\mathbf{u}} = \mathbf{h} + \sum_{i=1}^{n} \mathbf{f}_i = \mathbf{h} + \sum_{i=1}^{n} \mathbf{W}_i \lambda_i, \tag{3.8}$$

$$\dot{\mathbf{q}} = \mathbf{u}, \tag{3.9}$$

$$0 = \begin{cases} \mathbf{g}_i(\mathbf{q},t) & \text{displacement level} \\ \gamma_i = \mathbf{W}_i^\top \mathbf{u} + \zeta_i & \text{velocity level} \\ \dot{\gamma}_i = \mathbf{W}_i^\top \dot{\mathbf{u}} + \hat{\zeta}_i & \text{acceleration level} \end{cases} \tag{3.10}$$

The Lagrange-I formulation enforces the constraints (3.10) by generalized constraint forces $\mathbf{f}_i$. When the constraints are ideal, these generalized constraint forces have to be such that no virtual work is produced by arbitrary admissible virtual displacements. As a consequence, the generalized constraint forces must stand perpendicular to the manifolds $\mathbf{g}_i = 0$, $\gamma_i = 0$ and $\dot{\gamma}_i = 0$. Note that $\mathbf{W}_i = (\partial \mathbf{g}_i/\partial \mathbf{q})^\top$ is the associated differential to those manifolds, thus $\mathbf{f}_i = \mathbf{W}_i \boldsymbol{\lambda}_i$. In this context the matrix $\mathbf{W}_i$ is referred to as the generalized force direction of the i-th constraint. Furthermore, $\boldsymbol{\lambda}_i$ adresses the i-th constraint force. The equations of motion (3.8-3.9) and the constraints (3.10) form together a so-called differential algebraic equation (DAE), see also Sect. 2.5. The DAE (3.8-3.10) is of index one if all constraints appear on acceleration level. Index two is obtained if all constraints are formulated on velocity level. Constraints formulated on displacement level yield an index-3 DAE. The index of a DAE plays an important role in the numerical treatment of these kind of equations. In the following, the discussed concepts are extended to non-smooth dynamics, i.e. the constraints are replaced by unilateral contacts, friction and other non-smooth interactions of which the behaviour can not be described by equations anymore.

3.2 Non-Impulsive Motion

This section deals with set-valued force laws which model the behaviour of unilateral contacts, friction and other non-smooth interactions in the case of non-impulsive motion. The set-valued force laws are written as normal cone inclusions and are appended to the equations of motion as additional inclusions, similarly as bilateral constraints.

3.2.1 Set-Valued Force Laws

A set-valued force law can be interpreted as a constraint, whose constraint force $\boldsymbol{\lambda}_i$ is restricted to a convex set $\mathscr{C}_i$. Like (bilateral) constraints, a set-valued force law aims at enforcing a constraint equation between the states $\mathbf{q}$ and $\mathbf{u}$. Unlike (bilateral) constraints, a set-valued force law can not always enforce such a relation because the force $\boldsymbol{\lambda}_i$ is restricted. In the latter case the force $\boldsymbol{\lambda}_i$ becomes an impressed force. If an admissible force $\boldsymbol{\lambda}_i$ can be chosen such that a constraint equation on the states $\mathbf{q}$ and $\mathbf{u}$ is enforced, then the set-valued force law is in constraint mode. Otherwise, the force $\boldsymbol{\lambda}_i$ is impressed and the set-valued force law is in impressed mode.

Set-valued force laws can be expressed as the subdifferentials of the indicator function to different convex sets $\mathscr{C}_i$ [39, 41]. Similar to differential algebraic equations, it is distinguished between inclusions on displacement level (3.11), on velocity level (3.12) and on acceleration level (3.13), i.e.

$$-\mathbf{g}_i \in \partial \Psi_{C_i}(\boldsymbol{\lambda}_i) \equiv \mathscr{N}_{\mathscr{C}_i}(\boldsymbol{\lambda}_i), \tag{3.11}$$

$$\text{or } -\gamma_i \in \partial \Psi_{C_i}(\boldsymbol{\lambda}_i) \equiv \mathscr{N}_{\mathscr{C}_i}(\boldsymbol{\lambda}_i), \tag{3.12}$$

$$\text{or } -\dot{\gamma}_i \in \partial \Psi_{C_i}(\boldsymbol{\lambda}_i) \equiv \mathscr{N}_{\mathscr{C}_i}(\boldsymbol{\lambda}_i), \tag{3.13}$$

Table 3.1 Representation of a geometric set-valued force law on velocity level

$$\mathbf{g}_i = 0 \quad \Rightarrow -\gamma_i \in \mathscr{N}_{\mathscr{C}_i}(\lambda_i) \Rightarrow \begin{cases} \lambda_i \in \text{interior of } \mathscr{C}_i & \Rightarrow \text{constraint mode} \\ \lambda_i \in \text{boundary of } \mathscr{C}_i & \Rightarrow \text{impressed mode} \end{cases}$$

$$\mathbf{g}_i \neq 0 \quad \Rightarrow \text{impressed mode}$$

Geometric set-valued force laws are defined on displacement level by an inclusion of the form (3.11). This inclusion gathers two different states, a state at which λ_i is chosen within the set $\mathscr{C}_i$ to enforces the gap function $\mathbf{g}_i = 0$, and a state at which λ_i is at the boundary of the convex set $\mathscr{C}_i$, which yields $\mathbf{g}_i$ to be set-valued. In the first case, the set-valued force law is in constraint mode, as it enforces a constraint equation. In the second case, the set-valued force law is in impressed mode.

Example

Consider a unilateral contact i which prevents penetration between two bodies. The distance between the two bodies is characterized by the gap function g_i with $g_i > 0$ for an open, $g_i = 0$ for a closed and $g_i < 0$ for a penetrated contact. The unilateral contact can only transmit pressure forces, thus the set of admissible forces λ_i is $\mathscr{C}_i = \mathbb{R}_0^+$. The geometric set-valued force law which describes the behaviour of the unilateral contact yields

$$-g_i \in \mathscr{N}_{\mathbb{R}_0^+}(\lambda_i) \equiv \begin{cases} 0 & \text{if } \lambda_i > 0 \\ -\mathbb{R}_0^+ & \text{if } \lambda_i = 0 \end{cases} \tag{3.14}$$

Either the unilateral contact is closed, i.e. $g_i = 0$ and $\lambda_i \geq 0$, or the unilateral contact is open, i.e. $g_i > 0$ and $\lambda_i = 0$. In the first case, a positive force λ_i enforces the gap function g_i to be zero and the set-valued force law is in constraint mode. In the second case, the set-valued force law is in impressed mode, i.e. $\lambda_i = 0$.

A geometric set-valued force law can be expressed on velocity level by an inclusion of the form (3.12) if $\mathbf{g}_i = 0$, i.e. if we are on the manifold $\mathbf{g}_i = 0$, then the decision whether the set-valued force law is in constraint or in impressed mode can be done on velocity level. If $\mathbf{g}_i \neq 0$, then impressed mode applies in any case. Consult Table 3.1 for the representation of a geometric set-valued force law on velocity level. Take for instance a closed unilateral contact with $g_i = 0$. Such a contact can be described by the inclusion $-\gamma_i \in \mathscr{N}_{\mathbb{R}_0^+}(\lambda_i)$. Either the unilateral contact remains closed, that is a positive force λ_i enforces the relative velocity γ_i to be zero, or the unilateral contact is opening with a positive relative velocity $\gamma_i > 0$ and a force $\lambda_i = 0$. In the first case the associated geometric set-valued force law is in constraint mode, in the second case the set-valued force law is in impressed mode.

In the case $\mathbf{g}_i = 0$ and $\gamma_i = 0$, the geometric set-valued force law can be expressed on acceleration level by an inclusion of the form (3.13).

A kinematic set-valued force law, i.e. a set-valued force law which models friction, fails to have a description on displacement level. It can, however, always be characterized on velocity level by an inclusion of the form (3.12). In the case $\gamma_i = 0$, it can be characterized on acceleration level by an inclusion of the form (3.13).

3.2.2 *Examples of Set-Valued Force Laws*

In the following, set-valued force laws are given for the unilateral contact, planar and spatial friction, anisotropic friction and Coulomb-Contensou friction. Furthermore, it is discussed how bilateral constraints fit into the theory. All these different kinds of set-valued force laws differ from each other only by the set of admissible forces λ_i.

3.2.2.1 Geometric Set-Valued Force Law for a Unilateral Contact

The geometric set-valued force law associated with a unilateral contact uses a gap function g_i which defines the distance between two contact points. The unilateral contact is open for $g_i > 0$ and closed for $g_i = 0$. The set of admissible forces is $\mathbb{R}_0^+$. The behaviour of a closed stationary unilateral contact ($g_i = 0, \gamma_i = 0$) on acceleration level yields

$$-\dot{\gamma}_i \in \mathscr{N}_{\mathbb{R}_0^+}(\lambda_i) \equiv \begin{cases} 0 & \text{if } \lambda_i > 0 \\ -\mathbb{R}_0^+ & \text{if } \lambda_i = 0 \end{cases} \tag{3.15}$$

Either the closed stationary unilateral contact remains closed with a positive normal contact force λ_i and vanishing relative normal accelerations $\dot{\gamma}_i$, or the unilateral contact is opening, which requests positive relative normal accelerations $\dot{\gamma}_i$ and a vanishing normal contact force λ_i.

3.2.2.2 Kinematic Set-Valued Force Law for Planar Friction

A planar friction element is modeled by a kinematic set-valued force law, which connects a relative velocity γ_i in the friction zone with a friction force λ_i, which is restricted to the convex set $\mathscr{S}_p = \{\lambda_i \,|\, |\lambda_i| \le a_i\}$. The scalar a_i denotes the maximal friction force which is state-dependent in case of Coulomb friction, $a_i = \mu \lambda_j$. The force λ_j is the associated normal force which is either known or which is obtained from an associated force element j. The behaviour of a non-moving friction element ($\gamma_i = 0$) can be expressed on acceleration level to obtain

$$-\dot{\gamma}_i \in \mathscr{N}_{\mathscr{S}_p}(\lambda_i) \equiv \begin{cases} 0 & \text{if } |\lambda_i| < a_i \\ \mathbb{R}_0^+ \frac{\lambda_i}{|\lambda_i|} & \text{if } |\lambda_i| = a_i \end{cases} \tag{3.16}$$

Either the relative acceleration $\dot{\gamma}_i$ is zero and the friction element remains sticking with $|\lambda_i| < a_i$, or the friction element begins to slide, where the relative acceleration $\dot{\gamma}_i$ points in the opposite direction of the sliding force λ_i. Some frictional problems require the sliding force λ_i to depend on the relative velocities γ_i, i.e. Stribeck friction. In this case, the set-valued force law (3.16) is combined with an external force, which depends on the relative velocities γ_i. This external force has no set-valued character and its value is zero for $\gamma_i = 0$, see Sect. 7.6.2 or [41].

3.2.2.3 Frictional Unilateral Contact

Consider a frictional unilateral contact, i.e. a unilateral contact with friction. Such a contact is modeled by two force elements, by a unilateral contact and by a planar friction element. The two associated inclusions which express Coulomb friction can be seen in Fig. 3.1a. Note that the approach depicted in Fig. 3.1b does not allow for a displacement in the contact tangent plane without a separation. The associated friction model describes rough surfaces by infinite small peaks, which have to be overcome, see Fig. 3.1c. Sliding motion requests in this case separations, which is not a desired property in numerical simulations. Note that the separation time depends on the sliding velocity! Therefore, this work uses only the model depicted in Fig. 3.1a, which describes friction meaningful in a macroscopic way.

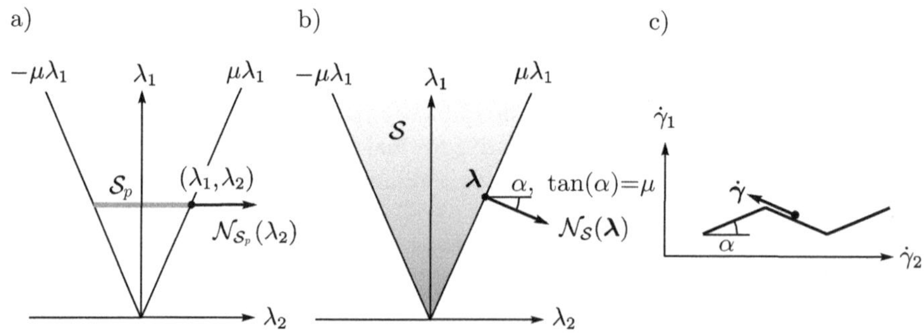

$$-\dot{\gamma}_1 \in \mathcal{N}_{\mathbb{R}_0^+}(\lambda_1)$$
$$-\dot{\gamma}_2 \in \mathcal{N}_{\mathcal{S}_p}(\lambda_2)$$
$$\mathcal{S}_p = \{\lambda_2 \mid |\lambda_2| \leq \mu\lambda_1\}$$

$$-\dot{\boldsymbol{\gamma}} \in \mathcal{N}_{\mathcal{S}}(\boldsymbol{\lambda}),\ \boldsymbol{\gamma} = (\dot{\gamma}_1\ \dot{\gamma}_2)^{\mathsf{T}},\ \boldsymbol{\lambda} = (\lambda_1\ \lambda_2)^{\mathsf{T}}$$
$$\mathcal{S} = \{(\lambda_1\ \lambda_2)^{\mathsf{T}} \mid \lambda_1 \geq 0, |\lambda_2| \leq \mu\lambda_1\}$$

Fig. 3.1 Different formulations for a frictional unilateral contact. Figure **a** shows the macroscopic interpretation, which is used in this work. The approach separates the frictional unilateral contact into a unilateral contact element and a friction element. Figures **b** and **c** depict a model which leads to separations

3.2.2.4 Kinematic Set-Valued Force Law for Spatial Friction

Different from planar friction, spatial friction requires a relative velocity γ_i and a friction force λ_i that are not scalars but planar vectors, which lie in the friction plane. Take for instance a point mass P on a plane. The position and velocity of the point are expressed in the inertial system I, i.e. $\mathbf{q} = _I\mathbf{r}_{OP}$ and $\mathbf{u} = \dot{\mathbf{q}}$. The plane is spanned by two orthogonal unit vectors $\mathbf{t}_1$ and $\mathbf{t}_2$. Both friction forces λ_i and relative velocities γ_i can now be defined in this local coordinate system, which yields the generalized force directions $\mathbf{W}_i$ to be $\mathbf{W}_i = ({_I}\mathbf{t}_1\ {_I}\mathbf{t}_2)$.

The friction forces λ_i of the spatial friction element are restricted to the convex set $\mathscr{S}_s = \{\lambda_i \in \mathbb{R}^2 \mid ||\lambda_i||_2 \leq a_i\}$, where a_i denotes the maximal friction force, which might also be state dependent. The inclusion for a non-moving frictional element ($\gamma_i = 0$) on acceleration level yields

$$-\dot{\gamma}_i \in \mathscr{N}_{\mathscr{S}_s}(\lambda_i) \equiv \begin{cases} 0 & \text{if } ||\lambda_i||_2 < a_i \\ \mathbb{R}_0^+ \frac{\lambda_i}{||\lambda_i||_2} & \text{if } ||\lambda_i||_2 = a_i \end{cases} \tag{3.17}$$

Either the relative acceleration γ_i is zero and the sticking friction element remains sticking with $||\lambda_i||_2 < a_i$, or the element begins to slide, whereas the relative acceleration γ_i points in the opposite direction to the sliding force λ_i.

3.2.2.5 Kinematic Set-Valued Force Law for Anisotropic Spatial Friction

As the (isotropic) spatial friction, the anisotropic spatial friction element requires a relative velocity γ_i and a friction force λ_i, which are defined by a local coordinate system in the friction plane. This coordinate system may be chosen such that the axes correspond to the maximal and minimal friction direction. In this case the set of admissible friction forces is of ellipse-shape, i.e.

$$\mathscr{S}_a = \{\lambda_i = (\lambda_x\, \lambda_y)^\top \in \mathbb{R}^2 \mid \left(\frac{\lambda_x}{c_i}\right)^2 + \left(\frac{\lambda_y}{d_i}\right)^2 \leq 1\}. \tag{3.18}$$

The inclusion which describes a non-moving anisotropic friction element ($\gamma_i = 0$) on acceleration level becomes

$$-\dot{\gamma}_i \in \mathscr{N}_{S_a}(\lambda_i). \tag{3.19}$$

3.2.2.6 Kinematic Set-Valued Force Law for Coulomb-Contensou Friction

Coulomb-Contensou friction describes the relation between drilling torque and classical Coulomb friction. Suppose for instance a rotating grinder machine and a wall. If the machine is rotating, then the relative velocities γ_i between wall and machine are mostly induced by the rotation and not by the movement along the wall.

Therefore, it is easier to move a switched-on grinder along a wall compared to a switched-off machine. Coulomb-Contensou friction is considered if the friction zone induces a drilling torque which affects the bodies movement. The friction force λ_i consist of a planar spatial friction force vector in the friction plane and a drilling moment perpendicular to it,

$$\lambda_i = \begin{pmatrix} \lambda_{t1} \\ \lambda_{t2} \\ \tau \end{pmatrix}. \tag{3.20}$$

The relative velocity γ_i is a planar vector for the movement in the friction plane and a rotation,

$$\gamma_i = \begin{pmatrix} \gamma_{t1} \\ \gamma_{t2} \\ \omega \end{pmatrix}. \tag{3.21}$$

A detailed description on Coulomb-Contensou friction considering different pressure force distributions in the friction zone is given in [43, 56]. A more simple model defines an ellipsoid shaped set of admissible friction forces λ_i,

$$\mathscr{S}_{cc} = \{\lambda_i \in \mathbb{R}^3 \mid \left(\frac{\lambda_x}{c_i}\right)^2 + \left(\frac{\lambda_y}{c_i}\right)^2 + \left(\frac{\lambda_z}{d_i}\right)^2 \leq 1\}, \tag{3.22}$$

in which the scalar c_i characterizes the Coulomb and d_i the Contensou friction, i.e. c_i corresponds to the absolute value of the maximal sticking force and d_i is associated with the maximal drilling torque. The description of a non-moving Coulomb-Contensou element ($\gamma_i = 0$) on acceleration level yields

$$-\dot{\gamma}_i \in \mathscr{N}_{S_{cc}}(\lambda_i). \tag{3.23}$$

3.2.2.7 Set-Valued Force Law for Bilateral Constraints

Bilateral constraints can be considered by choosing $\mathbb{R}^{m_i}$ as set of admissible forces λ_i. Thus every value of the force λ_i is admissible and the constraint is always enforced. In this sense, the bilateral constraint the most simple set-valued force law.

3.2.3 Equations of Motion in the Non-Smooth Case

In the following, different adequate formulations for the non-impulsive motion of a non-smooth mechanical system are shown [39, 41, 44]. The forces λ_i are incorporated into the equations of motion using a Lagrange-I formulation, as already done in Sect. 3.1 for the smooth case. The set-valued force laws are expressed on acceleration level as inclusions $-\lambda \in D(\dot{\gamma})$, which requests $\mathbf{g}_i = 0$ and $\gamma_i = 0$ for geometric and $\gamma_i = 0$ for kinematic set-valued force laws. Set-valued force laws for which the mentioned condition does not apply are in impressed mode and can explicitly be calculated from the state of the system. Kinematic transformations $\dot{\mathbf{u}} \to \dot{\gamma}$ are used to finally state the whole problem in relative entities, $\dot{\gamma} = \mathbf{G}\lambda + \mathbf{c}$, and to treat this equation together with the set-valued force laws.

The set-valued force laws are formulated on acceleration level and are linked to the equations of motion by a Lagrange-I formulation,

$$\mathbf{M}\dot{\mathbf{u}} = \mathbf{h} + \sum_{i=1}^{n} \mathbf{W}_i \lambda_i, \tag{3.24}$$

$$\dot{\mathbf{q}} = \mathbf{u}, \tag{3.25}$$

$$\dot{\gamma}_i = \mathbf{W}_i^\top \dot{\mathbf{u}} + \hat{\zeta}_i, \tag{3.26}$$

$$-\dot{\gamma}_i \in \partial\Psi_{C_i}(\lambda_i) \equiv \mathcal{N}_{\mathcal{C}_i}(\lambda_i) \quad \Leftrightarrow \quad \lambda_i \in \partial\Psi^\star_{C_i}(-\dot{\gamma}_i). \tag{3.27}$$

Note that several representations of the contact laws(3.27) exist. Elimination of the forces λ_i in the equations of motions (3.24) yields

$$\mathbf{M}\dot{\mathbf{u}} - \mathbf{h} - \sum_{i=1}^{n} \mathbf{W}_i \partial\Psi^\star_{C_i}\big(-\dot{\gamma}_i(\dot{\mathbf{u}})\big) \ni 0, \tag{3.28}$$

the integrated form of which is the associated optimization problem known as the extended principle of least constraints [41],

$$\min_{\dot{\mathbf{u}}} \frac{1}{2}(\dot{\mathbf{u}} - \dot{\mathbf{u}}_h)^\top \mathbf{M}(\dot{\mathbf{u}} - \dot{\mathbf{u}}_h) + \sum_{i=1}^{n} \Psi^\star_{Ci}\big(-\dot{\gamma}_i(\dot{\mathbf{u}})\big), \tag{3.29}$$

in which $\dot{\mathbf{u}}_h = \mathbf{M}^{-1}\mathbf{h}$ are the unconstrained accelerations of the system. Note that $\Psi^\star_{Ci}(-\dot{\gamma}_i)$ corresponds to the maximal possible rate of dissipation power in the i-th set-valued force law for $\dot{\gamma}_i$, i.e.

$$\Psi^\star_{Ci}(-\dot{\gamma}_i) = \sup_{\lambda_i}(-\lambda_i^\top \dot{\gamma}_i | \lambda_i \in \mathcal{C}_i). \tag{3.30}$$

Instead of using the equations of motions (3.24), it is also possible to transform the whole problem into relative coordinates to arrive at

$$\begin{aligned} \dot{\gamma} &= \mathbf{W}^\top \mathbf{M}^{-1} \mathbf{W}\lambda + \mathbf{W}^\top \mathbf{M}^{-1}\mathbf{h} + \hat{\zeta} =: \mathbf{G}\lambda + \mathbf{c}, \\ -\dot{\gamma}_i &\in \partial\Psi_{C_i}(\lambda_i), \\ \dot{\gamma} &= (\dot{\gamma}_1^\top \dots \dot{\gamma}_n^\top)^\top, \quad \lambda = (\lambda_1^\top \dots \lambda_n^\top)^\top, \\ \mathbf{W} &= (\mathbf{W}_1 \dots \mathbf{W}_n), \quad \hat{\zeta} = (\hat{\zeta}_1^\top \dots \hat{\zeta}_n^\top)^\top, \end{aligned} \tag{3.31}$$

where it has been taken into account that $\dot{\gamma}_i = \mathbf{W}_i^\top \dot{\mathbf{u}} + \hat{\zeta}_i$. Note that the so-called Delassus matrix $\mathbf{G}$ is positive definite if the force directions $\mathbf{W}_i$ are linearly independent, otherwise it is positive semidefinite. Furthermore, $\mathbf{G}$ is symmetric and all diagonal entries are larger than zero, as $\mathbf{M}$ is positive definite. Elimination of $\dot{\gamma}$ in (3.31) yields

$$\mathbf{G}\lambda+\mathbf{c}+\partial\Psi_{\mathscr{C}}(\lambda)\ni 0 \quad \Leftrightarrow \quad \begin{pmatrix} \sum_{j=1}^{n}\mathbf{G}_{1j}\lambda_j+\mathbf{c}_1+\partial\Psi_{\mathscr{C}_1}(\lambda_1) \\ \vdots \\ \sum_{j=1}^{n}\mathbf{G}_{nj}\lambda_j+\mathbf{c}_n+\partial\Psi_{\mathscr{C}_n}(\lambda_n) \end{pmatrix}\ni 0, \tag{3.32}$$

where $\mathbf{G}_{ij}=\mathbf{W}_i^\top\mathbf{M}\mathbf{W}_j$ denote the local Delassus matrices and $\mathbf{c}_i=\mathbf{W}_i^\top\mathbf{M}^{-1}\mathbf{h}+\hat{\zeta}_i$. Note that the problem can either be stated by one global inclusion or by n local inclusions which represent the n set-valued force laws. The associated optimization problem is

$$\min_{\lambda}\frac{1}{2}\lambda^\top\mathbf{G}\lambda+\lambda^\top\mathbf{c}+\Psi_{\mathscr{C}}(\lambda) \quad \Leftrightarrow \quad \min_{\lambda}\frac{1}{2}\lambda^\top\mathbf{G}\lambda+\lambda^\top\mathbf{c}+\sum_{i}^{n}\Psi_{\mathscr{C}_i}(\lambda_i), \tag{3.33}$$

which is the dual optimization problem to the extended principle of least constraints (3.29). Because $\Psi_{\mathscr{C}_i}(\lambda_i)=\infty$ for $\lambda_i\notin\mathscr{C}_i$, the optimization problem (3.33) can also be stated as a constrained optimization problem

$$\min_{\lambda\in\mathscr{C}}\frac{1}{2}\lambda^\top\mathbf{G}\lambda+\lambda^\top\mathbf{c} \quad \Leftrightarrow \quad \min_{\lambda_i\in\mathscr{C}_i}\frac{1}{2}\lambda^\top\mathbf{G}\lambda+\lambda^\top\mathbf{c}. \tag{3.34}$$

3.3 Impacts

In the case of an impact, the equations of motion have to be integrated over a singleton in time which yields the impact equations

$$\mathbf{M}(\mathbf{u}^+-\mathbf{u}^-)-\sum_{i=1}^{n}\mathbf{W}_i\Lambda_i=0. \tag{3.35}$$

Note that the pre and post-impact velocities are denoted by $\mathbf{u}^-$ and $\mathbf{u}^+$, and that the forces λ_i are replaced by the impulsive forces Λ_i, which are well defined in the case of an impact. In addition to the impact equations, the description of an impact requests an impact law. In this book Newton's impact law in inequality form is adopted, which can be written as

$$-(\gamma_i^+ +\varepsilon_i\gamma_i^-)\in\mathscr{N}_{\mathscr{D}_i}(\Lambda_i). \tag{3.36}$$

The scalar ε_i denotes the restitution coefficient, and the convex set $\mathscr{D}_i$ is the set of all admissible impulsive forces, whereas $\mathscr{D}_i$ might also be state-dependent in presence of friction. The impact law (3.36) is not only used for unilateral contacts but also for friction and any other kind of constitutive laws expressed as normal cone inclusions.

Example

Take for instance Newton's impact law in inequality form for a closed unilateral contact,

$$-(\gamma_i^+ + \varepsilon_i \gamma_i^-) \in \mathcal{N}_{\mathbb{R}_0^+}(\Lambda_i) \quad \Leftrightarrow \quad \Lambda_i \geq 0, \gamma_i^+ + \varepsilon \gamma_i^- \geq 0, \Lambda_i(\gamma_i^+ + \varepsilon \gamma_i^-) = 0. \tag{3.37}$$

The impact law accounts for two phenomena: An inversion of the relative velocity γ_i in the i-th unilateral contact due to a positive impulsive contact force Λ_i acting in the same contact, and an opening of a closed unilateral contact induced by an impulsive force Λ_i occurring in another contact of the mechanical system, take for instance the rocking rod [39].

Another impact law is Poisson's impact law [39], which works with a compression and an expansion phase, where the compression impulse Λ_i^C is related to the expansion impulse Λ_i^E by a restitution parameter ε_p. The compression phase corresponds to a complete inelastic impact, i.e. $-\gamma_i^C \in \mathcal{N}_{\mathcal{D}_i}(\Lambda_i^C)$. The expansion phase uses an expansion impulse Λ_i^E which is at least as large as $\varepsilon_p \Lambda^C$, which yields $-\gamma_i^E \in \mathcal{N}_{\mathcal{D}_i}(\Lambda_i^E - \varepsilon_p \Lambda_i^C)$. Note that Poisson's impact law requires the solution of two inclusion problems: one for the compression and one for the expansion phase. A detailed discussion on Poisson's impact law can be found in [39]. Both Newton's and Poisson's impact law do not hold for multi impact problems with remote action, take for instance the Newton cradle. Using a Frémond matrix ε instead of a scalar restitution coefficient ε, we are able to treat also multi impact problems with remote action by the impact law (3.36), see also Sect. 7.6.2. Detailed research on impact theory is done in [6, 7, 70].

3.4 Equality of Measures

Due to the presence of impulsive forces Λ, a non-smooth system can not be described solely by the equations of motions. Equalities of measures provide an elegant way to obtain a valid comprehensive description of a non-smooth system including the impact case. While the equations of motions use the accelerations $\dot{\mathbf{u}}$ and the forces $\boldsymbol{\lambda}$, which are not defined in the case of an impact, the description via equality of measures takes a measure $\mathrm{d}\mathbf{u}$ for the velocities and a measure $\mathrm{d}\mathbf{P}$ for the so-called percussions, which are well defined in the case of an impact.

In the following an example is discussed to introduce differential measures. The example intends to give the reader a short idea about measures. The reader is referred to the literature on distribution and measure theory, see for example [31, 64].

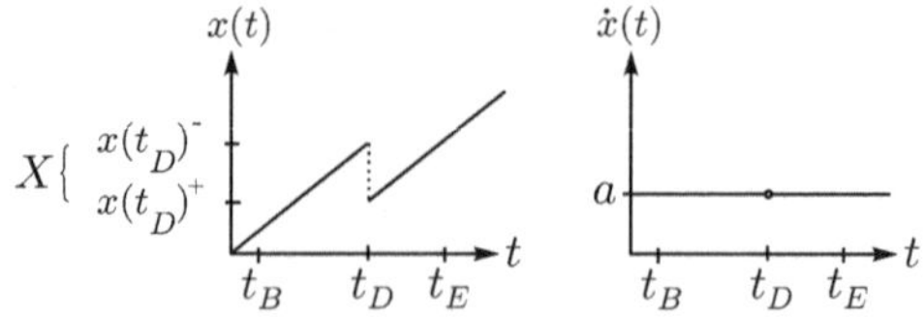

Fig. 3.2 Function $x(t)$ with discontinuity at the time t_D. The derivative $\dot{x}(t)$ of $x(t)$ is equal to a almost everywhere, except for a "single instantaneous infinite peak" at the time t_D

Example

Let $x(t)$ be the function depicted in figure 3.2. The function $x(t)$ has a discontinuity at the time t_D, where $t_B < t_D < t_E$. The derivative $\dot{x}(t)$ of $x(t)$ is equal to $a = const$ almost everywhere. At the discontinuity $t = t_D$, the derivative does not exist. At this time, the function $x(t)$ has a right and a left limit $x(t_D)^+$ and $x(t_D)^-$, and we set $X := x(t_D)^+ - x(t_D)^-$. We aim at determining the function value $x(t_E)$ for given a, X, and $x(t_B)$. A measure $\mathrm{d}x = \dot{x}\,\mathrm{d}t + (x^+ - x^-)\mathrm{d}\eta = a\,\mathrm{d}t + X\mathrm{d}\eta$ is defined which has a Lebesgue integrable term $a\,\mathrm{d}t$ and a purely atomic part $X\mathrm{d}\eta$. The term $\mathrm{d}\eta$ is a dirac point measure for the discontinuity at the time t_D, which means that the integral $\int_{\{t\}} \mathrm{d}\eta$ is equal to one if $t = t_D$ and zero otherwise. Calculation of $x(t_E)$ yields

$$\begin{aligned} x(t_E) &= x(t_B) + \int_{[t_B,t_E]} \mathrm{d}x = x(t_B) + \int_{t_B}^{t_E} a\,\mathrm{d}t + \int_{[t_B,t_E]} X\mathrm{d}\eta \\ &= x(t_B) + a(t_E - t_B) + X. \end{aligned} \tag{3.38}$$

The equations of motion and the impact equation can be combined in an equality of measures

$$\mathbf{M}\,\mathrm{d}\mathbf{u} - \mathbf{h}\,\mathrm{d}t - \sum_{i=1}^{n} \mathbf{W}_i\,\mathrm{d}\mathbf{P}_i = 0, \tag{3.39}$$

in which $\mathrm{d}\mathbf{u} = \dot{\mathbf{u}}\,\mathrm{d}t + (\mathbf{u}^+ - \mathbf{u}^-)\mathrm{d}\eta$ and $\mathrm{d}\mathbf{P}_i = \dot{\mathbf{P}}_i\mathrm{d}t + (\mathbf{P}_i^+ - \mathbf{P}_i^-)\mathrm{d}\eta = \lambda_i\,\mathrm{d}t + \Lambda_i\,\mathrm{d}\eta$ are differential measures for the velocity $\mathbf{u}$ and the percussion $\mathbf{P}$, respectively [94]. In the case of smooth motion, all measures $\mathrm{d}\eta$ vanish and a formal division by $\mathrm{d}t$ yields the equations of motion. In the case of an impact, the equality of measures (3.39) reduces to the impact equation.

In the following we try to combine the inclusion (3.36) and the inclusion (3.12) after integration in an inclusion

$$-\left((1+\varepsilon_i)\gamma_{Bi} + \int_{[t_B,t_E]} \mathrm{d}\gamma_i\right) \in \mathscr{N}_{\left(\int_{[t_B,t_E]} \mathrm{d}\mathscr{A}_i\right)}\left(\int_{[t_B,t_E]} \mathrm{d}\mathbf{P}_i\right), \tag{3.40}$$

in which $\mathrm{d}\gamma_i = \dot{\gamma}_i\,\mathrm{d}t + (\gamma_i^+ - \gamma_i^-)\mathrm{d}\eta$ is a measure for the relative velocity γ_i. The set of differential measures $\mathrm{d}\mathscr{A}_i = \mathscr{C}_i\,\mathrm{d}t + \mathscr{D}_i\,\mathrm{d}\eta$ is understood as

$$
\begin{aligned}
&\mathrm{d}\mathscr{A}_i = \{\mathrm{d}a_i \,|\, \mathrm{d}a_i = c_i\,\mathrm{d}t + d_i\,\mathrm{d}\eta,\, c_i \in \mathscr{C}_i,\, d_i \in \mathscr{D}_i\}, \\
&\textstyle\int \mathrm{d}\mathscr{A}_i := \left\{ \int \mathrm{d}a_i \,|\, a_i \in \mathscr{A} \right\}.
\end{aligned} \tag{3.41}
$$

This approach must be seen from an engineer's point of view and is only intended to give an idea on how set-valued laws for non-impulsive motion (3.12) and impulsive events (3.36) can be combined in a single formulation which helps to obtain Moreau's time-stepping scheme, see Sect. 5.3.2. The approach is not intended to be a mathematical proof of convergence and we are not sure if it acknowledges the complexity of the associated mathematics of measure differential inclusions in a complete manner. Of course much more sophisticated and detailed literature exist, see for example [63].

In the case of non-impulsive motion within a time interval $t_E - t_B = \Delta t$, the inclusion (3.40) reduces to

$$
-\left((1+\varepsilon_i)\gamma_{Bi} + \int_{[t_B,t_E]} \dot{\gamma}_i \mathrm{d}t\right) \in \mathscr{N}_{\left(\int_{[t_B,t_E]} \mathscr{C}_i \mathrm{d}t\right)} \left(\int_{[t_B,t_E]} \lambda_i \mathrm{d}t\right). \tag{3.42}
$$

Next, the mean value theorem for integration is used to obtain

$$
-\left((1+\varepsilon_i)\gamma_{Bi} + \int_{[t_B,t_E]} \dot{\gamma}_i \mathrm{d}t\right) \in \mathscr{N}_{\left(\mathscr{C}_i(t^\star)\Delta t\right)} \left(\lambda_i(t^{\star\star})\Delta t\right) \equiv \mathscr{N}_{\left(\mathscr{C}_i(t^\star)\right)} \left(\lambda_i(t^{\star\star})\right), \tag{3.43}
$$

in which $t^\star \in [t_B, t_E]$ and $t^{\star\star} \in [t_B, t_E]$. In the following we let t_E go versus t_B, i.e. $t_E \to t_B$, which yields $t^\star \to t_B$ and $t^{\star\star} \to t_B$. As $\int_{[t_B,t_E]} \dot{\gamma}_i \mathrm{d}t = 0$ for $t_E \to t_B$ we obtain

$$
-(1+\varepsilon_i)\gamma_{Bi} \in \mathscr{N}_{\left(\mathscr{C}_i(t_B)\right)} \left(\lambda_i(t_B)\right). \tag{3.44}
$$

Since multiplying the cone $\mathscr{N}_{\mathscr{C}_i}$ with the positive scalar $1/(1+\varepsilon_i)$ has no influence on the cone, inclusion (3.44) is equivalent to the inclusion (3.12). In the case of an impulsive event at t_B, the inclusion (3.40) reads

$$
-\left((1+\varepsilon_i)\gamma_{Bi}^- + \gamma_{Bi}^+ - \gamma_{Bi}^-\right) = -\left(\gamma_{Bi}^+ + \varepsilon_i \gamma_{Bi}^-\right) \in \mathscr{N}_{\mathscr{D}_i}(\Lambda_i), \tag{3.45}
$$

and becomes the inclusion stated in (3.36).

We may now define a geometric set-valued law which holds for both impulsive and non-impulsive motion ($t_E \to t_B$),

$$
\begin{aligned}
&g_i = 0 \quad \Rightarrow -\left((1+\varepsilon_i)\gamma_{Bi} + \textstyle\int_{[t_B,t_E]} \mathrm{d}\gamma_i\right) \in \mathscr{N}_{\left(\int_{[t_B,t_E]} \mathrm{d}\mathscr{A}_i\right)} \left(\textstyle\int_{[t_B,t_E]} \mathrm{d}\mathbf{P}_i\right) \\
&g_i \neq 0 \quad \Rightarrow \text{impressed mode}
\end{aligned} \tag{3.46}
$$

This law incorporates a set-valued force law expressed on velocity level and an impact law. A kinematic set-valued law which incorporates both impulsive and non-impulsive motion is given by the inclusion (3.44) without any further restrictions.

Remark

From a numerical point of view, the equality of measures (3.39) and the inclusion (3.40) can be interpreted as a discrete representation of a non-smooth system when $\Delta t \rightarrow 0$. As the discrete representation does not require the accelerations, the equality of measures (3.39) is also valid for the impact case. The inclusion (3.40) can be seen as a discrete law which reduces to either the set-valued force law (3.12) or the impact law (3.36) if $\Delta t \rightarrow 0$.

Chapter 4
Inclusion Problems

In this chapter it is discussed how inclusion problems can be solved. The method is demonstrated on the example of inclusions which describe the impact-free motion on acceleration level, i.e. inclusions which describe dynamical take-off or slip-stick transitions. As discussed in Sect. 3.2.3, one can either state n local inclusions for the n different set-valued force laws,

$$-\dot{\gamma}_i = -(\sum_{j=1}^{n} \mathbf{G}_{ij}\lambda_j + \mathbf{c}_i) \in \mathscr{N}_{\mathscr{C}_i}(\lambda_i), \quad i = 1 \ldots n, \tag{4.1}$$

or a global inclusion

$$-\dot{\gamma} = -(\mathbf{G}\lambda + \mathbf{c}) \in \mathscr{N}_{\mathscr{C}}(\lambda). \tag{4.2}$$

The corresponding global optimization problem is

$$\min_{\lambda} \lambda^{\top}\mathbf{G}\lambda + \lambda^{\top}\mathbf{c} + \Psi_{\mathscr{C}}(\lambda) \quad \Leftrightarrow \quad \min_{\lambda \in \mathscr{C}} \lambda^{\top}\mathbf{G}\lambda + \lambda^{\top}\mathbf{c}. \tag{4.3}$$

Other inclusion problems originating from the impact laws or from the later discussed time-stepping integration methods have the same structure as (4.1-4.3) and can therefore be treated in the same manner. In the following, different solution methods for inclusions problems of kind (4.1-4.3) are discussed. First, a very brief review on linear complementarity problems is given. Next, the augmented Lagrangian method is used to transform the inclusions into projective equations, which can be solved iteratively either by a Jacobi or Gauss-Seidel like technique. Note that the same projective equations can also be obtained by other approaches, i.e. uncoupling or exact regularization approaches, which are shortly discussed in Sect. 4.3. In this thesis, we refer to the approach which transforms inclusions into projective equations as augmented Lagrangian approach due to historical reasons, but we are aware that other notations exist. The chapter is based on the original publications [91, 93].

C. Studer: Numerics of Unilateral Contacts and Friction, LNACM 47, pp. 31–58.
springerlink.com

4.1 Linear Complementarity Problems

For simplicity, we restrict ourselves in this section to problems with n unilateral contacts,

$$-\dot{\gamma} = -(\mathbf{G}\boldsymbol{\lambda} + \mathbf{c}) \in \mathscr{N}_{\mathbb{R}_0^{n+}}(\boldsymbol{\lambda}). \tag{4.4}$$

Evaluation of (4.4) yields a complementarity between the relative accelerations $\dot{\gamma}$ and the forces $\boldsymbol{\lambda}$,

$$0 \preceq \dot{\gamma} = \mathbf{G}\boldsymbol{\lambda} + \mathbf{c} \perp \boldsymbol{\lambda} \succeq 0. \tag{4.5}$$

Possible methods to solve this linear complementarity problem are combinatoric methods, Lemke's method or iterative methods. Combinatoric methods just check every possible configuration and are therefore unsuited for larger problems. Lemke's method solves a linear complementarity problem by switching base vectors such that the complementarity is not affected. A possible instruction which solves (4.5) iteratively is given in [25, 67] to be

$$\lambda_i^{\nu+1} = \max\Big(-\frac{1}{G_{ii}}(\sum_{\substack{j=1\\ j\neq i}}^{n} G_{ij}\lambda_j^{\nu} + c_i),\, 0\Big). \tag{4.6}$$

This instruction corresponds with the treatment of unilateral contacts within the augmented Lagragian approach, which will be discussed in Sect. 4.2. The reader is referred to [25, 67] for a detailed discussion on linear complementarity problems. Note that the treatment of planar friction becomes more complicated, because the corresponding set-valued force law has to be decomposed into two set-valued force laws of the kind $-\gamma_i \in \mathscr{N}_{\mathbb{R}_0^+}(\lambda_i)$. The proper decomposition can be found in [39,44]. A set-valued force law for spatial friction becomes even more complicated, because the corresponding circular set $\mathscr{S}_s$ has to be linearized and approximated by a polygon to be treated within linear complementarity. Transforming inclusion problems in dynamics into linear complementarity problems gives a good insight into the structure of the problem. However, the approach is not very handy in application because the treatment of arbitrary set-valued force laws might become cumbersome. This work focuses on the augmented Lagrangian approach, which allows for a simple treatment of all kinds of set-valued force laws.

4.2 Augmented Lagrangian Approach

In this section it is shown how inclusion problems can be solved by using an augmented Lagrangian approach, which originates from optimization theory. The general problem is discussed, and an elegant way to solve a optimization problem subjected to inequality constraints of the form $\mathbf{k}(\mathbf{x}) \in \mathscr{C}$ is presented. The approach is applied to the optimization problem (4.3), which yields for each set-valued force law a local projective equation stated either in the forces λ_i or in the relative accelerations $\dot{\gamma}_i$. These projective equations are solved by Jacobi or Gauss-Seidel like iterative techniques. Two alternative approaches, which yield the same solution procedure, are discussed at the end of the section.

4.2.1 General Problem

Constrained optimization problems of strictly convex cost functions $f(\mathbf{x})$ can be solved by an augmented Lagrangian approach [15, 60, 75, 76]. Regarding a constrained optimization problem with affine equality constraints $\mathbf{k}(\mathbf{x}) = 0$,

$$\min_{\mathrm{x}} f(\mathbf{x}) \quad \text{for} \quad \mathbf{k}(\mathbf{x}) = 0, \tag{4.7}$$

the augmented Lagrangian function is [15]

$$L_a(\mathbf{x}, \mu) = f(\mathbf{x}) + \mu^\top \mathbf{k}(\mathbf{x}) + \frac{1}{2r} ||\mathbf{k}(\mathbf{x})||_2^2. \tag{4.8}$$

The vector μ are the Lagrange multipliers known from the ordinary Lagrangian, and r is an additional (inverse) penalty parameter. Finding the saddle point of the augmented Lagrangian function, i.e.

$$\min_{\mathrm{x}} \max_{\mu} L_a(\mathbf{x}, \mu), \tag{4.9}$$

is equivalent to solving the constrained optimization problem (4.7). In order to obtain a more general formulation, the penalty term $\frac{1}{2r} ||\mathbf{k}(\mathbf{x})||_2^2$ in the augmented Lagrangian function (4.8) is replaced by

$$\frac{1}{2} \mathbf{k}(\mathbf{x})^\top \mathbf{R} \mathbf{k}(\mathbf{x}) = \frac{1}{2} ||\mathbf{k}(\mathbf{x})||_{\mathrm{R}}^2 > 0 \quad \forall \mathbf{k}(\mathbf{x}) \tag{4.10}$$

with the matrix $\mathbf{R}$ being *symmetric* and *positive definite*, i.e. $\mathbf{x}^\top \mathbf{R} \mathbf{x} > 0 \, \forall \mathbf{x} \neq 0$. Choosing $\mathbf{R} = \frac{1}{r} \mathbf{I}$ yields again the penalty term in (4.8).

The augmented Lagrangian approach can also be used to solve constrained optimization problems with inequality constraints. Bertsekas [15], Luenberger [60] and Rockafellar [76] discuss the inequality constraints $k(\mathbf{x}) \geq 0$ and $a \leq k(\mathbf{x}) \leq b$. Let $\mathscr{C}$ be a convex set. In the following, general inequality constraints of the form $\mathbf{k}(\mathbf{x}) \in \mathscr{C}$ are considered, i.e. it is discussed how constrained optimization problems of the form

$$\min_{\mathrm{x}} f(\mathbf{x}) \quad \text{for} \quad \mathbf{k}(\mathbf{x}) \in \mathscr{C}, \tag{4.11}$$

can be approached. The restriction $\mathbf{k}(\mathbf{x}) \in \mathscr{C}$ in (4.11) is replaced by the restrictions $\mathbf{k}(\mathbf{x}) - \nu = 0$ and $\nu \in \mathscr{C}$. The saddle point problem associated with (4.11) is

$$\min_{\mathrm{x}, \nu \in \mathscr{C}} \max_{\mu} L_a(\mathbf{x}, \mu, \nu), \tag{4.12}$$

where L_a is the augmented Lagrangian function

$$L_a(\mathbf{x},\mu,\nu) = f(\mathbf{x}) + \mu^\top(\mathbf{k}(\mathbf{x}) - \nu) + \frac{1}{2}||\mathbf{k}(\mathbf{x}) - \nu||_\mathrm{R}^2. \tag{4.13}$$

Completing the square yields

$$L_a(\mathbf{x},\mu,\nu) = f(\mathbf{x}) - \frac{1}{2}\mu^\top\mathbf{R}^{-1}\mu + \frac{1}{2}||\mathbf{R}^{-1}\mu + \mathbf{k}(\mathbf{x}) - \nu||_\mathrm{R}^2. \tag{4.14}$$

The augmented Lagrangian function $L_a(\mathbf{x},\mu,\nu)$ is minimal with respect to $\nu \in \mathscr{C}$ if

$$\nu_0 = \operatorname*{argmin}_{\nu\in\mathscr{C}}\left(\frac{1}{2}||\mathbf{R}^{-1}\mu + \mathbf{k}(\mathbf{x}) - \nu||_\mathrm{R}^2\right) = \mathrm{prox}_\mathscr{C}^\mathrm{R}(\mathbf{R}^{-1}\mu + \mathbf{k}(\mathbf{x})). \tag{4.15}$$

Inserting ν_0 in (4.14) yields the augmented Lagrangian function

$$L_{a0}(\mathbf{x},\mu) = f(\mathbf{x}) - \frac{1}{2}\mu^\top\mathbf{R}^{-1}\mu + \frac{1}{2}||\mathrm{vdst}_\mathscr{C}^\mathrm{R}(\mathbf{R}^{-1}\mu + \mathbf{k}(\mathbf{x}))||_\mathrm{R}^2, \tag{4.16}$$

which is differentiable for all arguments. The solution of the saddle problem $\min_x \max_\mu L_{a0}(\mathbf{x},\mu)$ yields the conditions

$$\left(\frac{\partial L_{a0}}{\partial \mathbf{x}}\right)^\top = \left(\frac{\partial f(\mathbf{x})}{\partial \mathbf{x}}\right)^\top + \left(\frac{\partial \mathbf{k}(\mathbf{x})}{\partial \mathbf{x}}\right)^\top \mathbf{R}\,\mathrm{vdst}_\mathscr{C}^\mathrm{R}(\mathbf{R}^{-1}\mu + \mathbf{k}(\mathbf{x})) = 0,$$

$$\left(\frac{\partial L_{a0}}{\partial \mu}\right)^\top = \mathbf{R}^{-1}(-\mu + \mathbf{R}\,\mathrm{vdst}_\mathscr{C}^\mathrm{R}(\mathbf{R}^{-1}\mu + \mathbf{k}(\mathbf{x}))) = 0,$$

which can also be written as

$$0 = \left(\frac{\partial f(\mathbf{x})}{\partial \mathbf{x}}\right)^\top + \left(\frac{\partial \mathbf{k}(\mathbf{x})}{\partial \mathbf{x}}\right)^\top \mu, \tag{4.17}$$

$$\mu = \mathbf{R}\,\mathrm{vdst}_\mathscr{C}^\mathrm{R}(\mathbf{R}^{-1}\mu + \mathbf{k}(\mathbf{x})). \tag{4.18}$$

Using the definition of the $\mathrm{vdst}_\mathscr{C}^\mathrm{R}(\cdot)$ function, equation (4.18) can be simplified to obtain

$$\mathbf{k}(\mathbf{x}) = \mathrm{prox}_\mathscr{C}^\mathrm{R}(\mathbf{k}(\mathbf{x}) + \mathbf{R}^{-1}\mu). \tag{4.19}$$

Remark

Consider the relation (2.17), which links the inclusion $\mathbf{x} + \mathbf{R}^{-1}\mathscr{N}_\mathscr{C}(\mathbf{x}) \ni \xi$ to the projective equation $\mathbf{x} = \mathrm{prox}_\mathscr{C}^\mathrm{R}(\xi)$ without using an augmented Lagrangian approach. Note that the associated constrained optimization problem $\min_{x^\star\in\mathscr{C}} \frac{1}{2}||\mathbf{x}^\star - \xi||_\mathrm{R}^2$ corresponds to the definition of the $\mathrm{prox}_\mathscr{C}^\mathrm{R}(\xi)$ function. Therefore, the solution of the constrained optimization problem via augmented Lagrangian becomes unnecessary for this special case. Nevertheless, using $f(\mathbf{x}) = \frac{1}{2}||\mathbf{x} - \xi||_\mathrm{R}^2$ and $\mathbf{k}(\mathbf{x}) = \mathbf{x} \in \mathscr{C}$ within the augmented Lagrangian approach yields of course the same result,

$$\left.\begin{aligned} 0 &= \mathbf{R}(\mathbf{x}-\xi)+\mu \\ \mathbf{x} &= \mathrm{prox}_{\mathscr{C}}^{\mathrm{R}}(\mathbf{x}-\mathbf{R}^{-1}\mu) \end{aligned}\right\} \Rightarrow \mathbf{x} = \mathrm{prox}_{\mathscr{C}}^{\mathrm{R}}(\xi). \tag{4.20}$$

4.2.2 Application to Non-Smooth Dynamics

Applying the augmented Lagrangian to the optimization problem (4.3) yields

$$\dot{\gamma} = \mathbf{G}\lambda + \mathbf{c}, \tag{4.21}$$

$$-\dot{\gamma} = \mathbf{R}\,\mathrm{vdst}_{C}^{\mathrm{R}}(\lambda - \mathbf{R}^{-1}\,\dot{\gamma}), \tag{4.22}$$

in which the Lagrangian multipliers μ are replaced by to the negative accelerations $\dot{\gamma}$. Elimination of the accelerations $\dot{\gamma}$ yields a projective equation

$$\lambda = \mathrm{prox}_{\mathscr{C}}^{\mathrm{R}}(\lambda - \mathbf{R}^{-1}(\mathbf{G}\lambda + \mathbf{c})), \tag{4.23}$$

which we call λ-equation. Of course it is also possible to eliminate the forces λ instead of the relative accelerations $\dot{\gamma}$ which yields the $\dot{\gamma}$-equation

$$-\dot{\gamma} = \mathbf{R}\,\mathrm{vdst}_{\mathscr{C}}^{\mathrm{R}}(-\mathbf{R}^{-1}\dot{\gamma} + \mathbf{G}^{-1}(\dot{\gamma} - \mathbf{c}))). \tag{4.24}$$

In application it might be difficult to calculate $\mathrm{prox}_{\mathscr{C}}^{\mathrm{R}}(\cdot)$ with respect to an arbitrary positive definite symmetric matrix $\mathbf{R}$. In order to reduce complexity, $\mathbf{R}$ is chosen as block matrix according to (2.30), i.e. $\mathbf{R} = \mathbf{R}_1 \oplus \mathbf{R}_2 \oplus \ldots \oplus \mathbf{R}_n$, which allows for splitting the global projection $\mathrm{prox}_{\mathscr{C}}^{\mathrm{R}}$ on the set $\mathscr{C}$ into n local projections $\mathrm{prox}_{\mathscr{C}_i}^{\mathrm{R}_i}$ on the sets $\mathscr{C}_i$. Doing so the λ-equation (4.23) can be written as n λ_i-equations which represent the n set-valued force laws,

$$\lambda_i = \mathrm{prox}_{\mathscr{C}_i}^{\mathrm{R}_i}(\lambda_i - \mathbf{R}_i^{-1}(\sum_{j=1}^{n} \mathbf{G}_{ij}\lambda_j + \mathbf{c}_i)), \qquad i = 1\ldots n. \tag{4.25}$$

In the same manner, it is possible to express the global $\dot{\gamma}$-equation (4.24) as n individual $\dot{\gamma}_i$-equations

$$-\dot{\gamma}_i = \mathbf{R}_i\,\mathrm{vdst}_{\mathscr{C}_i}^{\mathrm{R}_i}\Big(-\mathbf{R}_i^{-1}\dot{\gamma}_i + \sum_{j=1}^{n}(\mathbf{G}^{-1})_{ij}\,(\dot{\gamma}_j - \mathbf{c}_j)\Big), \qquad i = 1\ldots n. \tag{4.26}$$

In a next step, the blocks $\mathbf{R}_i$ are chosen to be positive scaled identity matrices $\mathbf{I}_i$,

$$\mathbf{R}_{\mathrm{I}} := \frac{1}{r_1}\mathbf{I}_1 \oplus \frac{1}{r_2}\mathbf{I}_2 \oplus \ldots \oplus \frac{1}{r_n}\mathbf{I}_n, \quad r_i > 0, \qquad \mathbf{R}_{\mathrm{I}} \in \mathbb{R}^{m\times m},\ \mathbf{I}_i \in \mathbb{R}^{m_i\times m_i} \tag{4.27}$$

This choice allows not only to split the global λ and $\dot{\gamma}$ equations into local λ_i and $\dot{\gamma}_i$-equations but also to use the Euclidian norm for the projections $\mathrm{prox}_{\mathscr{C}_i}^{\mathrm{I}_i/r_i} = \mathrm{prox}_{\mathscr{C}_i}$. In this case the λ_i and the $\dot{\gamma}_i$-equations become

$$\lambda_i = \mathrm{prox}_{\mathscr{C}_i}(\lambda_i - r_i(\sum_{j=1}^{n} \mathbf{G}_{ij}\lambda_j + \mathbf{c}_i)), \qquad i = 1 \ldots n, \tag{4.28}$$

and

$$-\dot{\gamma}_i = \frac{1}{r_i}\,\mathrm{vdst}_{\mathscr{C}_i}\Big(-r_i\dot{\gamma}_i + \sum_{j=1}^{n}(\mathbf{G}^{-1})_{ij}\,(\dot{\gamma}_j - \mathbf{c}_j)\Big), \qquad i = 1 \ldots n. \tag{4.29}$$

The dynamics of the non-smooth system is now well described by either n λ_i-equations for the forces λ_i or n $\dot{\gamma}_i$-equations for the relative accelerations $\dot{\gamma}_i$. A mechanical system with a non-regular matrix $\mathbf{G}$ causes problems. The λ_i-equations of such a system might have non-unique solutions, whereas a description of such a system by $\dot{\gamma}_i$-equations is not even possible. Therefore only the λ_i-equations are discussed in the following. Some remarks concerning non-regular matrices $\mathbf{G}$ are given in Sect. 4.2.7.2.

4.2.3 *Iterative Solution Process*

We aim at solving a system of n projective λ_i-equations (4.28) using a Jacobi or Gauss-Seidel like iterative technique.

4.2.3.1 JORprox

A possible iterative scheme is

$$\lambda_i^{\nu+1} = \mathrm{prox}_{\mathscr{C}_i}(\lambda_i^{\nu} - r_i(\sum_{j=1}^{n} \mathbf{G}_{ij}\lambda_j^{\nu} + \mathbf{c}_i)), \qquad i = 1 \ldots n, \tag{4.30}$$

which we call the JORprox scheme. Note that this JORprox scheme corresponds to the Jacobi-relaxation (JOR) scheme [79]

$$\lambda_h^{\nu+1} = \lambda_h^{\nu} - \frac{\omega_h}{G_{hh}}(\sum_{k=1}^{m} G_{hk}\lambda_k^{\nu} + c_h), \qquad h = 1 \ldots m, \tag{4.31}$$

which converges to the solution of linear system $\mathbf{G}\lambda + \mathbf{c} = 0$ if $\mathbf{G} \in \mathbb{R}^{m,m}$ is a strictly diagonal dominant matrix,

$$\sum_{\substack{k=1 \\ k\neq h}}^{m} \left| \frac{G_{hk}}{G_{hh}} \right| < 1 \quad \forall h = 1 \ldots m. \tag{4.32}$$

The scalar ω_h is the relaxation parameter which must be chosen as $0 < \omega_h < 2$. The proposed scheme (4.30) performs one conventional Jacobi-relaxation step (4.31) in order to solve the linear system $\mathbf{G}\boldsymbol{\lambda} + \mathbf{c} = 0$ and adds a projection to ensure that the new update of the force $\lambda_i^{\nu+1}$ is within the set $\mathscr{C}_i$ of admissible forces. The scalar r_i represents the relaxation factor. Note that

$$r_i = \frac{\omega_h}{G_{hh}}, \tag{4.33}$$

for all rows h of $\mathbf{G}$ which belong to the i-th set-valued force law, i.e.

$$h = k + \sum_{j=1}^{i-1} m_j, \quad k = 1 \ldots m_i. \tag{4.34}$$

While a one-dimensional set-valued force law is represented by one row, a multi-dimensional set-valued force law addresses several rows in $\mathbf{G}$, according to the dimension of its associated $\mathbf{G}_{ii}$. It has to be pointed out that in the latter case the relaxation factors ω_h for the different rows belonging to the set-valued force law i can not be chosen independently due to (4.33).

Note that the JORprox scheme (4.30) can be written in global form as

$$\boldsymbol{\lambda}^{\nu+1} = \mathrm{prox}_{\mathscr{C}}(\boldsymbol{\lambda}^{\nu} - \mathbf{R}_{\mathrm{I}}^{-1}(\mathbf{G}\boldsymbol{\lambda}^{\nu} + \mathbf{c})), \tag{4.35}$$

where the matrix $\mathbf{R}_{\mathrm{I}}$ is chosen according to (4.27). This global form is used for convergence proofs in Sect. 4.2.4.

4.2.3.2 SORprox

It is also possible to solve the underlying linear system $\mathbf{G}\boldsymbol{\lambda} + \mathbf{c} = 0$ with a Gauss-Seidel-relaxation method (SOR) [79],

$$\lambda_h^{\nu+1} = \lambda_h^{\nu} - \frac{\omega_h}{G_{hh}} \Big(\sum_{k=1}^{h-1} G_{hk}\lambda_k^{\nu+1} + \sum_{k=h}^{m} G_{hk}\lambda_k^{\nu} + c_h\Big), \qquad h = 1 \ldots m, \tag{4.36}$$

which converges for positive definit matrices $\mathbf{G}$ if $0 < \omega_h < 2$. The corresponding SORprox scheme which solves the $\boldsymbol{\lambda}_i$-equations (4.25) yields

$$\boldsymbol{\lambda}_i^{\nu+1} = \mathrm{prox}_{\mathscr{C}_i}\Big(\boldsymbol{\lambda}_i^{\nu} - r_i\Big(\sum_{j=1}^{i-1} \mathbf{G}_{ij}\boldsymbol{\lambda}_j^{\nu+1} + \sum_{j=i}^{n} \mathbf{G}_{ij}\boldsymbol{\lambda}_j^{\nu} + \mathbf{c}_i\Big)\Big), \qquad i = 1 \ldots n. \tag{4.37}$$

JORprox

$$
\begin{aligned}
\boldsymbol{\lambda}_1^{\nu+1} &= \operatorname{prox}_{C_1}(\boldsymbol{\lambda}_1^{\nu} - r_1(\mathbf{G}_{11}\,\boldsymbol{\lambda}_1^{\nu} + \mathbf{G}_{12}\,\boldsymbol{\lambda}_2^{\nu} + \mathbf{G}_{13}\boldsymbol{\lambda}_3^{\nu} + \mathbf{c}_1))\\
\boldsymbol{\lambda}_2^{\nu+1} &= \operatorname{prox}_{C_2}(\boldsymbol{\lambda}_2^{\nu} - r_2(\mathbf{G}_{21}\,\boldsymbol{\lambda}_1^{\nu} + \mathbf{G}_{22}\,\boldsymbol{\lambda}_2^{\nu} + \mathbf{G}_{23}\boldsymbol{\lambda}_3^{\nu} + \mathbf{c}_2))\\
\boldsymbol{\lambda}_3^{\nu+1} &= \operatorname{prox}_{C_3}(\boldsymbol{\lambda}_3^{\nu} - r_3(\mathbf{G}_{31}\,\boldsymbol{\lambda}_1^{\nu} + \mathbf{G}_{32}\,\boldsymbol{\lambda}_2^{\nu} + \mathbf{G}_{33}\boldsymbol{\lambda}_3^{\nu} + \mathbf{c}_3))
\end{aligned}
$$

SORprox

$$
\begin{aligned}
\boldsymbol{\lambda}_1^{\nu+1} &= \operatorname{prox}_{C_1}(\boldsymbol{\lambda}_1^{\nu} - r_1(\mathbf{G}_{11}\,\boldsymbol{\lambda}_1^{\nu} + \mathbf{G}_{12}\,\boldsymbol{\lambda}_2^{\nu} + \mathbf{G}_{13}\boldsymbol{\lambda}_3^{\nu} + \mathbf{c}_1))\\
\boldsymbol{\lambda}_2^{\nu+1} &= \operatorname{prox}_{C_2}(\boldsymbol{\lambda}_2^{\nu} - r_2(\mathbf{G}_{21}\boldsymbol{\lambda}_1^{\nu+1} + \mathbf{G}_{22}\,\boldsymbol{\lambda}_2^{\nu} + \mathbf{G}_{23}\boldsymbol{\lambda}_3^{\nu} + \mathbf{c}_2))\\
\boldsymbol{\lambda}_3^{\nu+1} &= \operatorname{prox}_{C_3}(\boldsymbol{\lambda}_3^{\nu} - r_3(\mathbf{G}_{31}\boldsymbol{\lambda}_1^{\nu+1} + \mathbf{G}_{32}\boldsymbol{\lambda}_2^{\nu+1} + \mathbf{G}_{33}\boldsymbol{\lambda}_3^{\nu} + \mathbf{c}_3))
\end{aligned}
$$

Fig. 4.1 Visualization of the JORprox and SORprox iterative methods for a system which consists of three set-valued force laws. The updates $\lambda_i^{\nu+1}$ are highlighted to underline the difference between the two iterative methods

As in the case of the JORprox, the SORprox adds a projection to ensure that the update of the force $\lambda_i^{\nu+1}$ is within the set admissible forces $\mathscr{C}_i$. Different to the JORprox, the SORprox scheme makes use of the already calculated updates $\lambda_{j<i}^{\nu+1}$, see also Fig. 4.1. Consider in this context also the work of Jourdan et al. [49], who uses a non-linear block Gauss-Seidel type method. This method corresponds to a successive solution of the set-valued laws. Jourdan et al. write for each set-valued law an equation $\dot{\gamma}_i = \sum_{j=1}^{n} \mathbf{G}_{ij}\hat{\mathbf{P}}_j + \mathbf{c}_i$ and a projective equation $\hat{\mathbf{P}}_i = \operatorname{prox}_{\mathscr{C}_i}(\hat{\mathbf{P}}_i - r\dot{\gamma}_i)$. This two equations are solved for $\dot{\gamma}_i$ and $\hat{\mathbf{P}}_i$ under the assumption that all other relative accelerations $\dot{\gamma}_{j\neq i}$ and all other percussions $\hat{\mathbf{P}}_{j\neq i}$ are known. One-dimensional set-valued laws are solved straightforward, while multi-dimensional set-valued laws request a nested iterative approach. The solution is used for the calculation of the next set-valued law.

Remark

The SORprox can also be written in global form

$$\lambda^{\nu+1} = \operatorname{prox}_{\mathscr{C}}(\lambda^{\nu} - \mathbf{R}_{\mathrm{I}}^{-1}(\mathbf{L}\lambda^{\nu+1} + (\mathbf{U}+\mathbf{D})\lambda^{\nu} + \mathbf{c})), \quad \mathbf{G} = \mathbf{U}+\mathbf{D}+\mathbf{L}, \tag{4.38}$$

where $\mathbf{L}$ denotes a strictly lower triangular matrix, $\mathbf{D}$ a diagonal matrix and $\mathbf{U}$ a strictly upper triangular matrix. Different to the JORprox method, it is not obvious how $\lambda^{\nu+1}$ can be written as an explicit function of λ^{ν}.

In the following, we aim at finding an explicit formulation $\lambda^{\nu+1} = f(\lambda^{\nu})$ of an iterative scheme of the form

$$\lambda^{\nu+1} = \operatorname{prox}_{\mathscr{C}}(\mathbf{T}_1\lambda^{\nu+1} + \mathbf{T}_2\lambda^{\nu} + \mathbf{t}), \quad \mathbf{T}_1 + \mathbf{T}_2 = \mathbf{I} - \mathbf{R}_{\mathrm{I}}^{-1}\mathbf{G},\ \mathbf{t} = -\mathbf{R}_{\mathrm{I}}^{-1}\mathbf{c}. \tag{4.39}$$

Using the relation (2.17) it is possible to write (4.39) as

$$\lambda^{\nu+1} + \mathbf{R}_\mathrm{I}^{-1}\mathscr{N}_\mathscr{C}(\lambda^{\nu+1}) \ni \mathbf{T}_1\lambda^{\nu+1} + \mathbf{T}_2\lambda^{\nu} + \mathbf{t}. \tag{4.40}$$

Collecting the terms $\lambda^{\nu+1}$ yields

$$\lambda^{\nu+1} + (\mathbf{I}-\mathbf{T}_1)^{-1}\mathbf{R}_\mathrm{I}^{-1}\mathscr{N}_\mathscr{C}(\lambda^{\nu+1}) \ni (\mathbf{I}-\mathbf{T}_1)^{-1}(\mathbf{T}_2\lambda^{\nu} + \mathbf{t}). \tag{4.41}$$

If $(\mathbf{I}-\mathbf{T}_1)^{-1}\mathbf{R}_\mathrm{I}^{-1} = (\mathbf{R}_\mathrm{I} - \mathbf{R}_\mathrm{I}\mathbf{T}_1)^{-1}$ is symmetric and positive definite, then the inclusion (4.41) can again be written as projective equation

$$\lambda^{\nu+1} = \mathrm{prox}_\mathscr{C}^{\mathbf{R}_\mathrm{I}-\mathbf{R}_\mathrm{I}\mathbf{T}_1}\Big((\mathbf{I}-\mathbf{T}_1)^{-1}(\mathbf{T}_2\lambda^{\nu} + \mathbf{t})\Big). \tag{4.42}$$

Note the difference between (4.39) and (4.42). The instruction (4.39) contains forces $\lambda^{\nu+1}$ within the projective term and can be evaluated locally due to $\mathbf{R} = \mathbf{R}_\mathrm{I}$, see also equation (4.27). On the other hand, the projective equation (4.42) contains no forces $\lambda^{\nu+1}$ within the projective term and can only be evaluated globally, as $\mathbf{R} = \mathbf{R}_\mathrm{I} - \mathbf{R}_\mathrm{I}\mathbf{T}_1$ is in general not block diagonal according (2.30). Regarding the SORprox scheme, the matrix $\mathbf{R} = \mathbf{R}_\mathrm{I} - \mathbf{R}_\mathrm{I}\mathbf{T}_1$ becomes $\mathbf{R}_\mathrm{I} + \mathbf{L}$, which is at least positive semi-definite for some choices of $\mathbf{R}_\mathrm{I}$ but not symmetric. Nevertheless, a skewed global projection behaviour can be observed, see Fig. 4.10. Probably the symmetric SOR method, i.e. the SSOR method, might bring further advancements, consider for example [51].

4.2.4 *Choice of r_i and Convergence*

This section starts with a general discussion on Banach's fixed point theorem. The theorem is afterwards applied to the JORprox method, and convergence is discussed. Solely projective equations which originates from normal cone inclusions of associated type are considered, i.e. Coulomb friction is excluded. For such associated normal cone inclusions an optimization problem can be stated, and the convex set $\mathscr{C}_i$ of admissible forces λ_i is not state-dependent. Regarding the JORprox scheme, convergence can only be guaranteed in the case of $\mathbf{G}$ being strictly diagonally dominant. Furthermore, the choice of r_i is discussed. Some remarks about the convergence of the SORprox method are given at the end of this section.

4.2.4.1 Banach's Fixed Point Theorem

Banach's fixed point theorem states that an iterative instruction of the form

$$\xi^{\nu+1} = \mathbf{F}(\xi^{\nu}), \quad \xi \in \mathbb{R}^m, \quad \mathbf{F}: \mathbb{R}^m \to \mathbb{R}^m \tag{4.43}$$

converges to a unique solution ξ of $\xi = \mathbf{F}(\xi)$ if

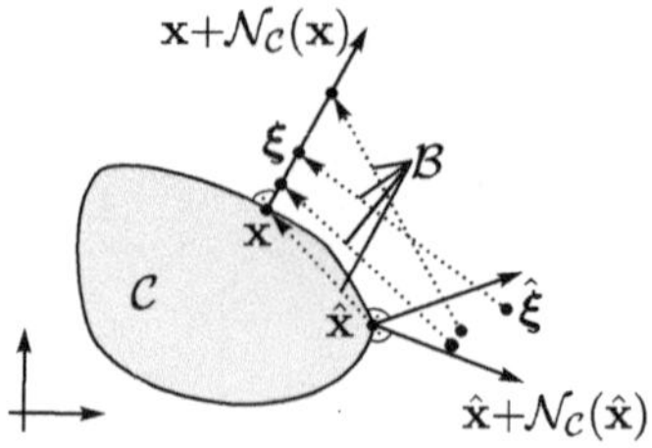

Fig. 4.2 Set $\mathscr{B}$ for $\mathbf{R}$ being the identity matrix $\mathbf{I}$. Some vectors which connect the set of points $\hat{\mathbf{x}} + \mathscr{N}_{\mathscr{C}}(\hat{\mathbf{x}})$ with the set of points $\mathbf{x} + \mathscr{N}_{\mathscr{C}}(\mathbf{x})$ are depicted as *dotted arrows*. The set $\mathscr{B}$ consist of the lengths $||\cdot||_{\mathrm{I}}^2$ of all such vectors

$$||\mathbf{F}(\xi) - \mathbf{F}(\hat{\xi})|| \leq L||\xi - \hat{\xi}||, \quad L < 1, \quad \forall \xi, \hat{\xi} \in \mathbb{R}^m. \tag{4.44}$$

A small Lipschitz constant L results in a better convergence. Arbitrary norms $||\cdot||$ can be used. For more detailed information see for example [79].

4.2.4.2 Contractivity of the $\mathrm{prox}_{\mathscr{C}}^{\mathrm{R}}$ Function

We prove the contractivity of the $\mathrm{prox}_{\mathscr{C}}^{\mathrm{R}}$ with respect to $||\cdot||_{\mathrm{R}}$.

Theorem

The $\mathrm{prox}_{\mathscr{C}}^{\mathrm{R}}$ mapping is a contraction with respect to $||\cdot||_{\mathrm{R}}$, i.e.

$$||\mathbf{x} - \hat{\mathbf{x}}||_{\mathrm{R}}^2 := ||\mathrm{prox}_{\mathscr{C}}^{\mathrm{R}}(\xi) - \mathrm{prox}_{\mathscr{C}}^{\mathrm{R}}(\hat{\xi})||_{\mathrm{R}}^2 \leq ||\xi - \hat{\xi}||_{\mathrm{R}}^2 \quad \forall \xi, \hat{\xi}, \tag{4.45}$$

if the set $\mathscr{C}$ is not state-dependent.

Proof

Using the relation (2.17), i.e. $\xi \in \mathbf{x} + \mathscr{N}_{\mathscr{C}}(\mathbf{x}) \Leftrightarrow \mathbf{x} = \mathrm{prox}_{\mathscr{C}}(\xi)$, the right-hand side of (4.45) can be written as

$$||\xi - \hat{\xi}||_{\mathrm{R}}^2 \in ||\mathbf{x} + \mathbf{R}^{-1}\mathbf{y} - \hat{\mathbf{x}} - \mathbf{R}^{-1}\hat{\mathbf{y}}||_{\mathrm{R}}^2 := \mathscr{B}, \qquad \begin{array}{l} \mathbf{y} \in \mathscr{N}_{\mathscr{C}}(\mathbf{x}), \hat{\mathbf{y}} \in \mathscr{N}_{\mathscr{C}}(\hat{\mathbf{x}}), \\ \forall \mathbf{x} \in \mathscr{C}, \forall \hat{\mathbf{x}} \in \mathscr{C}. \end{array} \tag{4.46}$$

The set $\mathscr{B}$ contains the lengths $||\cdot||_{\mathrm{R}}^2$ of all vectors which connect the set of points $\hat{\mathbf{x}} + \mathbf{R}^{-1}\mathscr{N}_{\mathscr{C}}(\hat{\mathbf{x}})$ with the set of points $\mathbf{x} + \mathbf{R}^{-1}\mathscr{N}_{\mathscr{C}}(\mathbf{x})$, see also Fig. 4.2. Both the distances $||\xi - \hat{\xi}||_{\mathrm{R}}^2$ and $||\mathbf{x} - \hat{\mathbf{x}}||_{\mathrm{R}}^2$ are elements of $\mathscr{B}$. In the following, we show that $||\mathbf{x} - \hat{\mathbf{x}}||_{\mathrm{R}}^2$ is the smallest element of $\mathscr{B}$. We evaluate $||\cdot||_{\mathrm{R}}^2$ in (4.46) to obtain

$$\begin{aligned}\mathscr{B} &\equiv \Big((\mathbf{x}-\hat{\mathbf{x}})+\mathbf{R}^{-1}(\mathbf{y}-\hat{\mathbf{y}})\Big)^{\top}\mathbf{R}\Big((\mathbf{x}-\hat{\mathbf{x}})+\mathbf{R}^{-1}(\mathbf{y}-\hat{\mathbf{y}})\Big)\\ &\equiv (\mathbf{x}-\hat{\mathbf{x}})^{\top}\mathbf{R}(\mathbf{x}-\hat{\mathbf{x}})+2\cdot(\mathbf{y}-\hat{\mathbf{y}})^{\top}(\mathbf{x}-\hat{\mathbf{x}})+(\mathbf{y}-\hat{\mathbf{y}})^{\top}\mathbf{R}^{-1}\mathbf{R}\mathbf{R}^{-1}(\mathbf{y}-\hat{\mathbf{y}})\\ &\equiv ||\mathbf{x}-\hat{\mathbf{x}}||_{\mathrm{R}}^2+||\mathbf{R}^{-1}(\mathbf{y}-\hat{\mathbf{y}})||_{\mathrm{R}}^2-2\cdot\mathbf{y}^{\top}(\hat{\mathbf{x}}-\mathbf{x})-2\cdot\hat{\mathbf{y}}^{\top}(\mathbf{x}-\hat{\mathbf{x}}).\end{aligned} \tag{4.47}$$

Consulting the definition of the normal cone (2.10), i.e.

$$\mathscr{N}_{\mathscr{C}}(\mathbf{x}) \equiv \{\mathbf{y}\,|\,\mathbf{y}^{\top}(\mathbf{x}^{\star}-\mathbf{x})\leq 0\},\ \forall\mathbf{x}^{\star}\in\mathscr{C},\mathbf{x}\in\mathscr{C}, \tag{4.48}$$

yields $\mathbf{y}^{\top}(\hat{\mathbf{x}}-\mathbf{x})\leq 0$ and $\hat{\mathbf{y}}^{\top}(\mathbf{x}-\hat{\mathbf{x}})\leq 0$. As a consequence, the summands in (4.47) are positive. As a normal cone always contains the zero element, $\mathbf{y}$ and $\hat{\mathbf{y}}$ can be chosen equal to zero, which yields $||\mathbf{x}-\hat{\mathbf{x}}||_{\mathrm{R}}^2$ to be the smallest element of $\mathscr{B}$. Furthermore, it holds

$$||\xi-\hat{\xi}||_{\mathrm{R}}^2 \geq ||\mathbf{x}-\hat{\mathbf{x}}||_{\mathrm{R}}^2 = ||\mathrm{prox}_{\mathscr{C}}^{\mathrm{R}}(\xi)-\mathrm{prox}_{\mathscr{C}}^{\mathrm{R}}(\hat{\xi})||_{\mathrm{R}}^2, \tag{4.49}$$

which completes the proof. In the following we use $\mathbf{R}=\mathbf{I}$, which yields

$$||\mathrm{prox}_{\mathscr{C}}(\xi)-\mathrm{prox}_{\mathscr{C}}(\hat{\xi})||_2 \leq ||\xi-\hat{\xi}||_2 \quad \forall\xi,\hat{\xi}. \tag{4.50}$$

4.2.4.3 Application of Banach's Fixed Point Theorem on the JORprox Method

This section shows that the JORprox method converges if the underlying JOR method does, where the proof is restricted to associated problems, i.e. to problems for which the set $\mathscr{C}$ does not depend on the forces λ. Regarding the JORprox instruction (4.35), it can be distinguished between the JOR instruction $\mathbf{F}(\lambda)$ and the additional projection,

$$\mathbf{F}(\lambda) = \lambda - \mathbf{R}_I^{-1}(\mathbf{G}\lambda+\mathbf{c}) := \mathbf{T}_{\mathrm{JOR}}\lambda+\mathbf{t}_{\mathrm{JOR}}, \quad \lambda^{\nu+1} = \mathrm{prox}_{\mathscr{C}}(\mathbf{F}(\lambda^{\nu})), \tag{4.51}$$

with $\mathbf{T}_{\mathrm{JOR}} = (\mathbf{I}-\mathbf{R}_{\mathrm{I}}^{-1}\mathbf{G})$ and $\mathbf{t}_{\mathrm{JOR}} = -\mathbf{R}^{-1}\mathbf{c}$ being the iterative matrix and the constant vector of the classical Jacobi relaxation scheme. Next, the contractivity of the $\mathrm{prox}_{\mathscr{C}}$ function can be used to obtain

$$||\mathrm{prox}_{\mathscr{C}}(\mathbf{F}(\lambda))-\mathrm{prox}_{\mathscr{C}}(\mathbf{F}(\hat{\lambda}))||_2 \leq ||\mathbf{F}(\lambda)-\mathbf{F}(\hat{\lambda})||_2 \leq ||\mathbf{T}_{\mathrm{JOR}}||_2\,||\lambda-\hat{\lambda}||_2. \tag{4.52}$$

The right-hand side of (4.52) is the convergence criterion of the classical JOR method evaluated in the spectrum norm. Note that the matrix $\mathbf{T}_{\mathrm{JOR}}$ is symmetric and that therefore the spectrum norm $||\cdot||_2$ provides the minimal matrix norm $||\cdot||_{\star}$,

$$||\mathbf{T}_{\mathrm{JOR}}|| \geq ||\mathbf{T}_{\mathrm{JOR}}||_{\star} = ||\mathbf{T}_{\mathrm{JOR}}||_2 = \rho, \tag{4.53}$$

see also Sect. 2.1. Thus the JORprox method converges if the underlying JOR method does.

In the following, the convergence of the JORprox method is investigated more closely. The method converges if the Lipschitz constant $L = ||\mathbf{T}_{\mathrm{JOR}}||_2$ is smaller than one. This criterion can be evaluated for a global choice of $r_i = r \forall i$, see also [32]. Let μ_{G} and $\mu_{\mathrm{T_{JOR}}}$ be vectors which contain the eigenvalues of $\mathbf{G}$ and $\mathbf{T}_{\mathrm{JOR}}$, respectively. Note that $\mu_{\mathrm{G}} \succeq 0$ because $\mathbf{G}$ is at least positive semidefinite. The eigenvalues of the two matrices are related by

$$\mu_{\mathrm{T_{JOR}}} = 1 - r\mu_{\mathrm{G}}, \tag{4.54}$$

which allows for writing the Lipschitz constant as a function of μ_{G},

$$L = \rho(\mathbf{T}_{\mathrm{JOR}}) = \max(|1 - r\max(\mu_{\mathrm{G}})|, |1 - r\min(\mu_{\mathrm{G}})|). \tag{4.55}$$

The Lipschitz constant is minimal for

$$1 - r\max(\mu_{\mathrm{G}}) = 1 - r\min(\mu_{\mathrm{G}}) \quad \Rightarrow \quad r = \frac{2}{\max(\mu_{\mathrm{G}}) + \min(\mu_{\mathrm{G}})}. \tag{4.56}$$

Note that if the matrix $\mathbf{G}$ is only positive semidefinite, then at least one eigenvalue is equal to zero. In this case the Lipschitz constant is at least one and convergence can not be proven.

The calculation of the eigenvalues μ_{G} can be omitted when estimating the spectrum norm $||\mathbf{T}_{\mathrm{JOR}}||_2$ by the rowsum norm $||\mathbf{T}_{\mathrm{JOR}}||_\infty$,

$$L = ||\mathbf{T}_{\mathrm{JOR}}||_2 \leq \underbrace{||\mathbf{T}_{\mathrm{JOR}}||_\infty}_{\tilde{L}} = ||\mathbf{I} - \mathbf{R}_{\mathrm{I}}^{-1}\mathbf{G}||_\infty = \max_h \left(|1 - \omega_h| + \omega_h \sum_{\substack{k=1 \\ k \neq h}}^{m} \left| \frac{G_{hk}}{G_{hh}} \right| \right), \tag{4.57}$$

where ω_h is related to r_i by equation (4.33). Note that the estimation (4.57) requests that $\mathbf{T}_{\mathrm{JOR}}$ is symmetric, which yields $||\mathbf{T}_{\mathrm{JOR}}||_2$ to be the minimal matrix norm, see (4.53). In Fig. 4.3, the estimation $\tilde{L}$ of the Lipschitz constant L is shown as a function of ω_h, together with the two summands and that build it up, see also (4.57). The left diagram corresponds to a matrix $\mathbf{G}$ which is strictly diagonal dominant. In this case, a minimum Lipschitz constant $\tilde{L} < 1$ is reached for $\omega_h = 1$, and

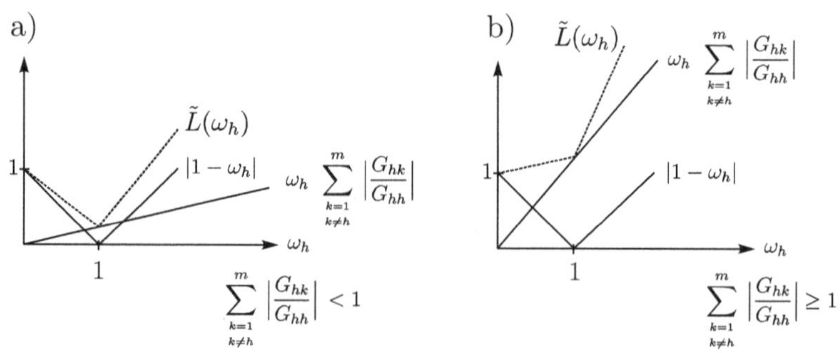

Fig. 4.3 Minimizing the Lipschitz constant $\tilde{L}(\omega_h)$ (4.57). In Fig. **a**, **G** is strictly diagonal dominant, and a minimum can be reached for $\omega_h = 1$. In Fig. **b**, **G** is not strictly diagonal dominant, and the Lipschitz constant $\tilde{L}(\omega_h)$ is always larger than one [91]

convergence can be guaranteed. In the right diagram, the matrix $\mathbf{G}$ is not strictly diagonal dominant. As a consequence, the estimate $\tilde{L}$ of the Lipschitz constant continuously increases from one. Although convergence can not be guaranteed for this case, the numerical scheme might still work. This results from the fact that $\tilde{L}(\omega_h)$ is close to one for small values of ω_h. However, choosing ω_h small does *not* solve the problem, because this choice influences the stop criterion of the algorithm, which is based on a relative and absolute error estimate: The choice $\omega_h = 0$ causes $r_i = 0$ by (4.27). As a consequence, the iterative instruction (4.23) for inclusion i becomes independent from all other inclusions j, and the algorithm terminates already after the first loop, because after it the values of λ_i are not changed anymore. Note that the parameter r_i controls the influence of the j-th inclusion on the i-th inclusion.

Based on these observations, the following choice of the parameter r_i is suggested. Note again the structure of the Delassus matrix $\mathbf{G}$, i.e.

$$\mathbf{G} = \begin{pmatrix} \mathbf{G}_{11} & \dots & \mathbf{G}_{1n} \\ \vdots & \ddots & \vdots \\ \mathbf{G}_{n1} & \dots & \mathbf{G}_{nn} \end{pmatrix}, \qquad \mathbf{G}_{ij} = \mathbf{W}_i^{\top} \mathbf{M}^{-1} \mathbf{W}_j, \tag{4.58}$$

see also (3.31). The dimensions of the local Delassus matrices $\mathbf{G}_{ij}$ depends on the dimensions m_i and m_j of the associated set-valued force laws i and j, i.e. $\mathbf{G}_{ij} \in \mathbf{R}^{m_i \times m_j}$. The local Delassus matrices $\mathbf{G}_{ii}$ are square matrices with dimension m_i, i.e. the local Delassus matrix $\mathbf{G}_{ii}$ of a one-dimensional set-valued force law corresponds to the entry G_{hh} of $\mathbf{G}$, where h denotes the row in $\mathbf{G}$ which belongs to the set-valued force law i.

First, the case of a one-dimensional set-valued force law i is discussed. If the matrix $\mathbf{G}$ is strictly diagonal dominant, then it is proposed to use

$$r_i = \frac{1}{G_{hh}}, \tag{4.59}$$

in which h denotes the row in $\mathbf{G}$ which belongs to the set-valued force law i. If $\mathbf{G}$ is not strictly diagonal dominant, then r_i has to be chosen small. A good empiric criterion is

$$r_i = \frac{1}{\sum\limits_{k=1}^{m} |G_{hk}|}. \tag{4.60}$$

If the diagonal elements predominate, then criteria (4.59) and (4.60) become similar. Another possibility is to scale all the values r_i obtained by (4.59) with a scalar $0 < \alpha \leq 1$. A multi-dimensional set-valued force law i addresses more than one row in $\mathbf{G}$, as already discussed in Sect. 4.2.3. In general, the criteria (4.59) and (4.60) are still applicable, where h addresses any of the associated rows in $\mathbf{G}$, preferably the one which minimizes r_i. It is also possible to use (4.59) with the average value of all G_{hh} which are associated with the set-valued force law i.

Remark

Convergence problems might be encountered when the values of $\mathbf{G}_{hh}$ are of very different magnitudes. In this case, also the values of ω_h are of very different magnitudes, see equation (4.33). As a consequence, convergence problems might occur. This behaviour has been observed for set-valued force laws in which the single components of the corresponding force $\boldsymbol{\lambda}_i$ have very different influence on the mechanical behaviour, for example the drilling torque within the Coulomb-Contensou friction. In this case it is advisable to scale the corresponding components of the force $\boldsymbol{\lambda}_i$, which affects the corresponding rows in $\mathbf{W}_i$ and therefore also the values G_{hh}. Take for instance a two-dimensional set-valued force law, of which the second component λ_2 of the force $\boldsymbol{\lambda}_i$ is scaled by a scalar α,

$$\mathbf{f}_i = \mathbf{W}_i\boldsymbol{\lambda}_i = \begin{pmatrix}\mathbf{w}_1 & \alpha\mathbf{w}_2\end{pmatrix}\begin{pmatrix}\lambda_1 \\ \lambda_2/\alpha\end{pmatrix} \quad \Rightarrow \quad \mathbf{G}_{ii} = \begin{pmatrix}\mathbf{w}_1^\top\mathbf{w}_1 & \alpha\mathbf{w}_1^\top\mathbf{w}_2 \\ \alpha\mathbf{w}_2^\top\mathbf{w}_1 & \alpha^2\mathbf{w}_2^\top\mathbf{w}_2\end{pmatrix}. \tag{4.61}$$

Note that scaling the components of the force $\boldsymbol{\lambda}_i$ also requests scaling the set of admissible forces $\mathscr{C}_i$. Note the link to (2.21). The following three approaches are equivalent: working with arbitrary choices of positive definite symmetric matrices $\mathbf{R}$, working with transformed convex sets $\mathscr{C}$, and transforming the forces $\boldsymbol{\lambda}_i$ by altering the generalized force directions $\mathbf{W}_i$.

4.2.4.4 Convergence the SORprox Method

It is presumable that the convergence of the SORprox method is the same as of the classical SOR method, i.e. it converges for positive definite matrices $\mathbf{G}$. However, a proof is still missing. A main difficulty stems from the lack of a global representation of the form $\boldsymbol{\lambda}^{\nu+1} = \mathrm{prox}_{\mathscr{C}}^{\mathrm{R}}(\mathbf{F}(\boldsymbol{\lambda}^{\nu}))$. The SORprox scheme has shown satisfactory behaviour in practical application and should be used as the preferred iterative solution method. Note that dependent generalized force directions $\mathbf{W}_i$ cause $\mathbf{G}$ to be positive semidefinite. In this case the Lipschitz constant of the classical Gauss-Seidel relaxation method can reach the value one, which yields the iteration to become stationary in some directions. This corresponds to the fact that non-unique solutions must be expected. Note that the Lipschitz constant is never larger than one because $\mathbf{G}$ is at least positive semidefinite.

4.2.5 *Use of Non-Diagonal* R*'s*

In principle it is possible to assign any positive definite symmetric matrix to $\mathbf{R}$. So far we restricted ourselves to $\mathbf{R} = \mathbf{R}_{\mathrm{I}}$ as defined in (4.27), because this choice allows for evaluating the $\mathrm{prox}_{\mathscr{C}}^{\mathrm{R_I}}$ by the Euclidian norm. The choice of a non diagonal $\mathbf{R}$ results in extensive projections, and has therefore so far no practical significance.

Nevertheless, two choices of $\mathbf{R}$, which give a better insight into the problem, are discussed.

4.2.5.1 Choosing $\mathbf{R} = \mathbf{G}$, where $\mathbf{G}$ is Positive Definite

When choosing $\mathbf{R} = \mathbf{G}$, the equation (4.23) can be solved straightforwardly within one projection,

$$\lambda = \text{prox}_{\mathscr{C}}^{\mathrm{G}}(\lambda - \mathbf{G}^{-1}(\mathbf{G}\lambda + \mathbf{c}) = \text{prox}_{\mathscr{C}}^{\mathrm{G}}(-\mathbf{G}^{-1}\mathbf{c}). \tag{4.62}$$

At a first glance, this choice simplifies the problem, because no iterative solution is required anymore. The difficulty arises when calculating $\text{prox}_{\mathscr{C}}^{\mathrm{G}}$. Note that the projection to the global set $\mathscr{C}$ can no longer be split into projections to the local sets $\mathscr{C}_i$, because $\mathbf{G}$ is in general not of the form (2.30), and that therefore all edges of $\mathscr{C}$ caused by the cartesian product of the local sets $\mathscr{C}_i$ must be considered separately. This procedure becomes equivalent to the combinatorial solution of the non-smooth problem. Take for instance a mechanical system with two unilateral contacts, thus $\mathbf{G} \in \mathbb{R}^{2\times 2}$, $\mathbf{c} \in \mathbb{R}^2$, $\mathscr{C} = \mathbb{R}_0^{2+}$. In order calculate $\text{prox}_{\mathbb{R}_0^{2+}}^{\mathrm{G}}$, the projection of a point ξ_0 to a straight line $\mathbf{x}_1 + t\mathbf{v}$ with respect to $||\cdot||_{\mathrm{G}}$ is examined, see Fig. 4.4a. Note that the level lines $f(\xi) = ||\xi - \xi_0||_{\mathrm{G}} = const$ are ellipses. We denote by $\mathbf{s}$ all vectors emanating form ξ_0 to any point on $\mathbf{x}_1 + t\mathbf{v}$ and choose $\mathbf{v}_\perp \perp \mathbf{v}$ in a way that $\mathbf{s} = \mathbf{v}_\perp + \alpha\mathbf{v}$. In order to calculate the shortest vector $\mathbf{s}_0$ with respect to $||\cdot||_{\mathrm{G}}$, the term $||\mathbf{s}||_{\mathrm{G}}^2 = \mathbf{s}^\top\mathbf{G}\mathbf{s}$ is minimized with respect to α,

$$\left(\frac{\partial}{\partial\alpha}(\mathbf{s}^\top\mathbf{G}\mathbf{s})\right)^\top = \mathbf{v}^\top\mathbf{G}(\mathbf{v}_\perp + \alpha_0\mathbf{v}) = 0 \quad \Rightarrow \quad \alpha_0 = -\frac{\mathbf{v}^\top\mathbf{G}\mathbf{v}_\perp}{\mathbf{v}^\top\mathbf{G}\mathbf{v}}, \quad \mathbf{s}_0 = \mathbf{v}_\perp + \alpha_0\mathbf{v}. \tag{4.63}$$

Doing so for $\mathbf{v} = (1\,0)^\top$ yields the direction of $\mathbf{s}_0$ to become $(-1 \quad G_{11}/G_{12})^\top$. For $\mathbf{v} = (0\,1)^\top$ the direction $(-1 \quad G_{21}/G_{22})^\top$ is obtained. With these two results it is

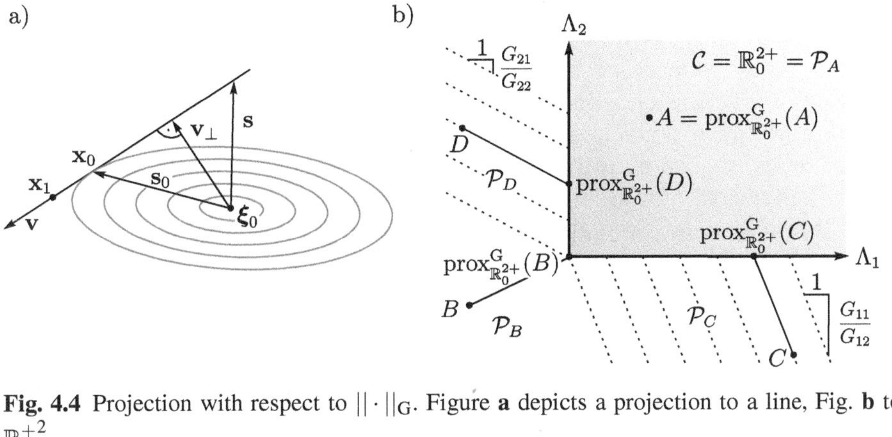

Fig. 4.4 Projection with respect to $||\cdot||_{\mathrm{G}}$. Figure **a** depicts a projection to a line, Fig. **b** to $\mathbb{R}_0^{+2}$

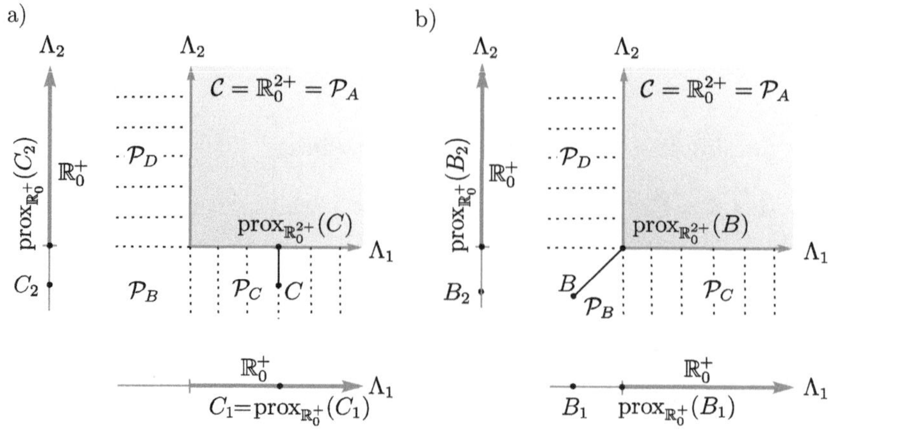

Fig. 4.5 Projection with respect to $||\cdot||_{\mathbf{R}_I}$. The global projection to $\mathbb{R}_0^{+2}$ can be split in two local projections to $\mathbb{R}_0^+$. Figure **a** depicts the projection of point C, Fig. **b** the projection of point B

possible to calculate $\text{prox}^{\mathbf{G}}_{\mathbb{R}_0^{2+}}$, as shown in Fig. 4.4b. Note the four different areas $\mathscr{P}_A$, $\mathscr{P}_B$, $\mathscr{P}_C$ and $\mathscr{P}_D$, which correspond to the different configurations of the set-valued force laws. If $\mathbf{G}^{-1}\mathbf{c} \in \mathscr{P}_A$ or $\mathbf{G}^{-1}\mathbf{c} \in \mathscr{P}_B$, then both unilateral contacts are closed or open, respectively. If $\mathbf{G}^{-1}\mathbf{c} \in \mathscr{P}_C$, then the unilateral contact 2 opens, if $\mathbf{G}^{-1}\mathbf{c} \in \mathscr{P}_D$, then the unilateral contact 1 opens. Calculating $\text{prox}^{\mathbf{G}}_{\mathbf{R}_0^{2+}}$ requests these four distinctions, and the approach can therefore be viewed as a combinatoric solution method. Note the difference to the projection with respect to the Euclidian norm, which can be split into local projections to $\mathbb{R}_0^+$, see Fig. 4.5. While the projection to $\mathbb{R}_0^{2+}$ distinguishes between the four sets $\mathscr{P}_A$ to $\mathscr{P}_D$, the local projections to $\mathbb{R}_0^+$ only check wether Λ_1 and Λ_2 are in $\mathbb{R}_0^+$ or not. The advantages of splitting the global set $\mathscr{C}$ into local sets $\mathscr{C}_i$ become obvious for systems with more than two set-valued force laws. Consider, for example, a mechanical system with n unilateral contacts. The associated global set $\mathscr{C}$ has 2^n "surfaces and edges" which correspond to the 2^n different states of the set-valued force laws. When performing a projection on this global set $\mathscr{C}$, 2^n different cases have to be considered. When splitting the global set $\mathscr{C}$ into n local sets $\mathscr{C}_i$, an iteration is required which requests n decisions per step. To conclude, choosing $\mathbf{R} = \mathbf{G}$ allows for solving the non-smooth problem by one projection to a global set $\mathscr{C}$. This projection can not be split into projections to local sets $\mathscr{C}_i$ and therefore the 2^n different configurations have to be considered. If $\mathbf{R}$ is chosen such that the global projection to $\mathscr{C}$ can be replaced by local projections to $\mathscr{C}_i$, then only n decisions are required. Note that in this case the solution of the non-smooth problem is not obtained within one projection step, but the problem has to be solved iteratively.

As the number of configurations explodes for a rising number of set-valued force laws, the choice $\mathbf{R} = \mathbf{G}$ is so far useless in practical application. Nevertheless, the observations made might be helpful for the construction of more advanced iterative solution methods. It would also be interesting to see whether Lemke's algorithm for solving linear complementarity problems can be interpreted by a special choice of $\mathbf{R}$. In addition, one might think of replacing the skewed projection with respect to $||\cdot||_{\mathrm{G}}$ by Euclidian projections to a transformed set $\mathscr{C}$. Doing so one may apply perhaps sophisticated projection algorithms which come from computer graphics science.

4.2.5.2 Choosing $\mathbf{R}_i = \mathbf{G}_{ii}$, where $\mathbf{G}_{ii}$ is Positive Definite

A one-dimensional set-valued force law i with $m_i = 1$ can be solved straightforward by the inclusion (4.28) if $r_i = 1/G_{ii}$ and if all other forces $\lambda_{j\neq i}$ are known. In general, this does not hold anymore for multi-dimensional set-valued force laws i with $m_i > 1$. Choosing $\mathbf{R}_i = \mathbf{G}_{ii} = \mathbf{W}_i^\top \mathbf{M}^{-1}\mathbf{W}_i$ solves this problem. Note that a local representation of (4.28) is still possible, i.e. for each set-valued force law an inclusion of the form

$$\lambda_i = \mathrm{prox}_{\mathscr{C}_i}^{\mathrm{G}_{ii}}(\lambda_i - \mathbf{G}_{ii}^{-1}(\sum_{j=1}^{n} \mathbf{G}_{ij}\lambda_j + \mathbf{c}_i)) = \mathrm{prox}_{\mathscr{C}_i}^{\mathrm{G}_{ii}}(-\mathbf{G}_{ii}^{-1}(\sum_{\substack{j=1\\ j\neq i}}^{n} \mathbf{G}_{ij}\lambda_j + \mathbf{c}_i)) \quad (4.64)$$

can be stated. These inclusions can be solved iteratively by a JORprox or SORprox like method. Unlike the discussed JORprox and SORprox methods, an iterative scheme based on (4.64) would solve multi-dimensional set-valued force laws within one iteration step for given forces $\lambda_{j\neq i}$. Note that the choice of $\mathbf{R}_i = \mathbf{G}_{ii}$ requests the inversion of $\mathbf{G}_{ii}$. Furthermore, the projection to the local set $\mathscr{C}_i$ might be cumbersome. Take for instance spatial friction. Let the corresponding $\mathrm{prox}_{\mathscr{S}_{s,i}}^{\mathrm{G}_{ii}}$ function performs a projection with respect to $||\cdot||_{\mathrm{G}_{ii}}$ onto a circular set $\mathscr{S}_{s,i} = \{\mathbf{x} \in \mathbb{R}^2 \,|\, ||\mathbf{x}||_2 \leq 1\}$, see also Fig. 4.6. Let $\mathbf{x}$ denote all points lying on the unit circle. Finding the point $\mathbf{x}_0$ on the circle which is closest to a an arbitrary point ξ_0 with respect to $||\cdot||_{\mathrm{G}_{ii}}$ is then equivalent to solving the optimization problem $\min_x ||\xi_0 - \mathbf{x}||_{\mathrm{G}_{ii}}$ under the restriction $\mathbf{x}^\top\mathbf{x} = 1$. The corresponding saddle-point problem yields

$$\min_{\xi}\max_{\mu}\,(\xi_0 - \mathbf{x})^\top \mathbf{G}_{ii}(\xi_0 - \mathbf{x}) + \mu\,(\mathbf{x}^\top\mathbf{x} - 1), \quad (4.65)$$

in which μ is a Lagrangian multiplier. Solving the saddle point problem (4.65) yields a fourth-order polynomial which has to be solved for the the Lagrange multiplier μ_0,

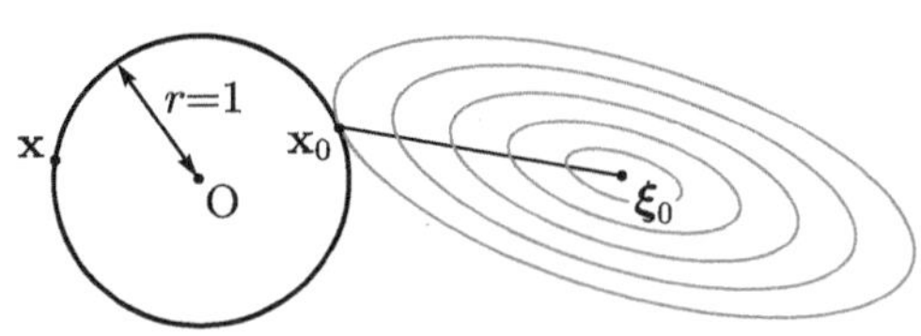

Fig. 4.6 Projection to a unit circle with respect to $||\cdot||_{\mathrm{G}_{ii}}$

$$\left((\mathbf{G}_{ii}+\mu_0\mathbf{I})^{-1}\mathbf{G}_{ii}\xi_0\right)^{\top}\left((\mathbf{G}_{ii}+\mu_0\mathbf{I})^{-1}\mathbf{G}_{ii}\xi_0\right)=1. \tag{4.66}$$

Finally the point $\mathbf{x}_0$ which lies closest to ξ_0 follows from

$$\mathbf{x}_0=(\mathbf{G}_{ii}+\mu_0\mathbf{I})^{-1}\mathbf{G}_{ii}\xi_0. \tag{4.67}$$

To conclude, choosing $\mathbf{R}_i=\mathbf{G}_{ii}$ allows for solving the i-th set-valued force law straightforward if the other forces $\lambda_{j\neq i}$ are known. In this case the $\mathrm{prox}_{\mathscr{C}_i}^{\mathbf{G}_{ii}}$ function has to be evaluated with respect to the norm $||\cdot||_{\mathbf{G}_{ii}}$ which might become cumbersome. In addition, the matrix $\mathbf{G}_{ii}$ has to be inverted. Therefore, it is still advised to use the classical JORprox and SORprox scheme discussed in Sect. 4.2.3.

Remark

As already discussed, the choice of an arbitrary positive definite symmetric $\mathbf{R}_i$ affects the projection to the set $\mathscr{C}_i$. In the case of a circular set $\mathscr{S}_{s,i}$ this is not a desired property, because the Euclidian projection to this set is trivial. In the case of a elliptical set $\mathscr{S}_{a,i}$ on which no simple Euclidian projection exist, such a matrix $\mathbf{R}_i$ may simplify the projection procedure. Working with arbitrary positive definite symmetric matrices $\mathbf{R}$ is equivalent to working with Euclidian projections to transformed sets, see also Sect. 2.3.2.

4.2.6 Successive Solution of the Set-Valued Force Laws

In the previous Sect. 4.2.5.2 it has been shown that the choice $\mathbf{R}_i=\mathbf{G}_{ii}$ solves each set-valued force law i straightforward if all other forces $\lambda_{j\neq i}$ are known in advance. The algorithm assumes that the forces are λ^{ν} and calculates for each set-valued force law a new force $\lambda^{\nu+1}$. This leads to a JORprox like approach

$$\lambda_i^{\nu+1}=\mathrm{prox}_{\mathscr{C}_i}^{\mathbf{G}_{ii}}\Big(-\mathbf{G}_{ii}^{-1}\big(\sum_{\substack{j=1\\ j\neq i}}^{n}\mathbf{G}_{ij}\lambda_j^{\nu}+\mathbf{c}_i\big)\Big). \tag{4.68}$$

This does not consider the fact that already updated forces $\lambda_{j<i}^{\nu+1}$ are available when calculating $\lambda_i^{\nu+1}$. Taking these updated forces $\lambda_{j<i}^{\nu+1}$ into account yields a SORprox like scheme

$$\lambda_i^{\nu+1}=\mathrm{prox}_{\mathscr{C}_i}^{\mathbf{G}_{ii}}\Big(-\mathbf{G}_{ii}^{-1}\big(\sum_{j=1}^{i-1}\mathbf{G}_{ij}\lambda_j^{\nu+1}+\sum_{j=i+1}^{n}\mathbf{G}_{ij}\lambda_j^{\nu}+\mathbf{c}_i\big)\Big). \tag{4.69}$$

This procedure corresponds to a successive Gauss-Seidel like solution of the set-valued force laws [65, 49]. Note that the classical SORprox scheme uses the choice

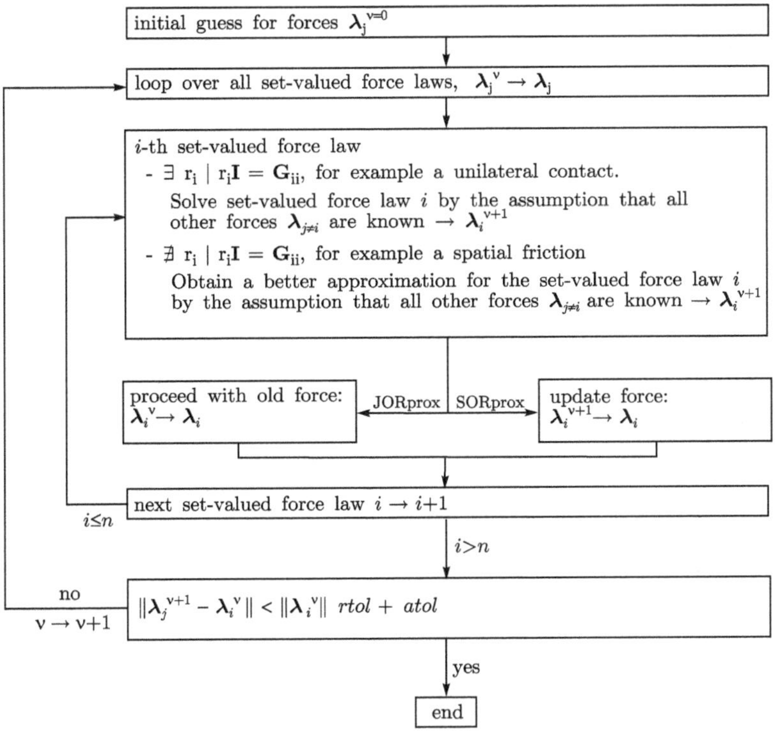

Fig. 4.7 Interpretation of the SORprox and JORprox method as successive solution of the set-valued force laws [91]

$\mathbf{R}_i = r_i\,\mathbf{I}$. If r_i is chosen in a way that $r_i\mathbf{I} = \mathbf{G}_{ii}$, then the corresponding SORprox scheme solves the i-th set-valued force law straightforward for known $\lambda_{j\neq i}$. Note that such a choice for r_i is in general not possible for multi-dimensional set-valued force laws $m_i > 1$. In this case only a better approximation of the force λ_i is obtained. In this sense also the classical SORprox method can be viewed as a successive solution of the set-valued force laws, see Fig. 4.7.

4.2.7 Examples

In the following, two examples are discussed. The first example visualizes the JORprox and the SORprox iteration process, the second example focusses on a mechanical system with a non-regular matrix $\mathbf{G}$.

4.2.7.1 Visualization of the JORprox and SORprox Scheme

Consider the mechanical system depicted in Fig. 4.8. The system consists of a bar with mass m and inertia Θ. The bar lies on two bearings which are modeled as unilateral contacts. Furthermore, a vertical force F_y and a moment F_φ act on the bar's center of mass S. The positive definite matrix $\mathbf{G}$ and the vector $\mathbf{c}$ are

$$\mathbf{G} = \frac{1}{m\Theta}\begin{pmatrix} \Theta + ma^2 & \Theta + mab \\ \Theta + mab & \Theta + mb^2 \end{pmatrix}, \qquad \mathbf{c} = \frac{1}{m\Theta}\begin{pmatrix} \Theta F_y + ma F_\varphi \\ \Theta F_y + mb F_\varphi \end{pmatrix}, \tag{4.70}$$

The corresponding set of admissible forces is $\mathscr{C} = \mathbb{R}_0^{2+}$. The parameters are $a = 0.5$ m, $b = 0.1$ m, $m = 1$ kg and $\Theta = 1/12\,\mathrm{kg\,m}^2$. This example aims at visualizing

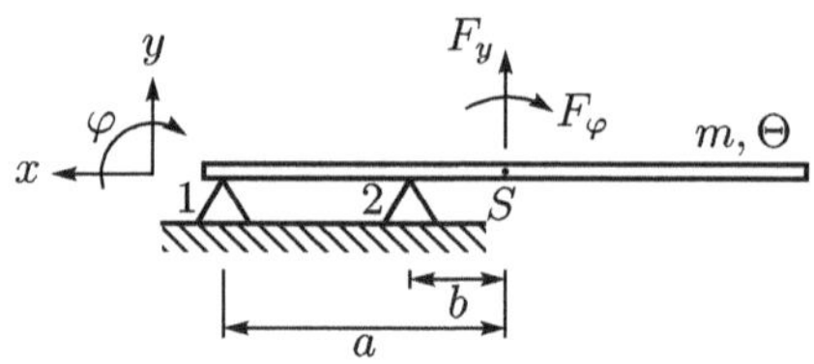

Fig. 4.8 Mechanical system

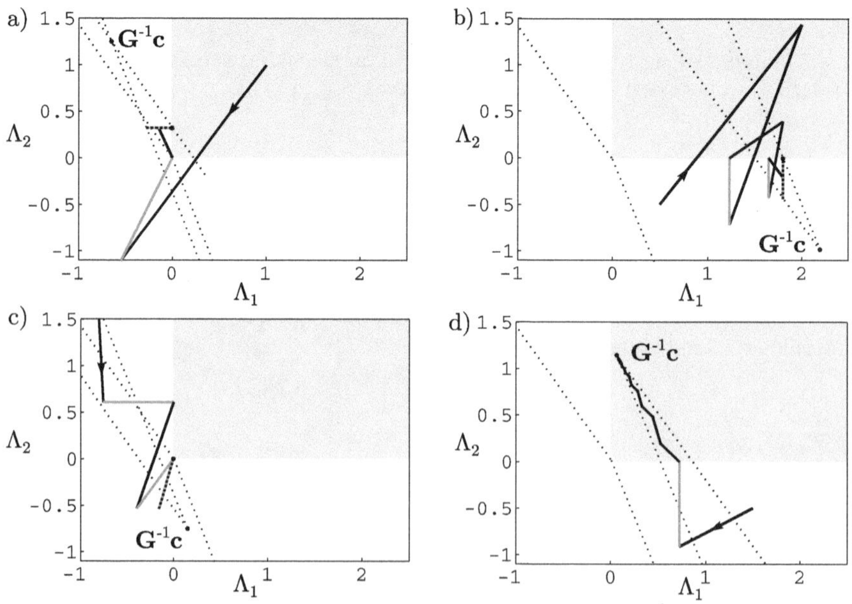

Fig. 4.9 Visualization of the JORprox scheme

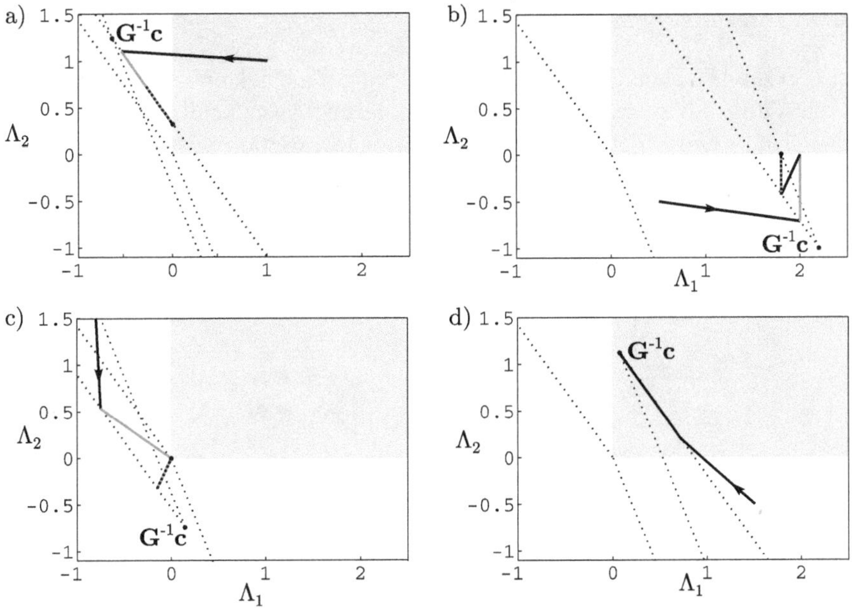

Fig. 4.10 Visualization of the SORprox scheme

the convergence of the JORprox and SORprox iteration method in the (Λ_1,Λ_2)-space. Figures 4.9a-d and 4.10a-d show the convergence of the JORprox and SOR-prox method for different loadings F_y and F_φ. The classical JOR and SOR steps, i.e. $\Lambda^{\nu+1} = \mathbf{T}_{\mathrm{JOR/SOR}}\Lambda^{\nu} + \mathbf{t}_{\mathrm{JOR/SOR}}$, are indicated by a solid black line. The additional projection $\Lambda^{\nu+1} = \mathrm{prox}_{\mathscr{C}}(\mathbf{F}_{\mathrm{JOR/SOR}}(\Lambda^{\nu},\Lambda^{\nu+1}))$ is depicted as grey line. The thin black dotted lines indicate the directions of projection to the Λ_1- and Λ_2-axis with respect to $||\cdot||_{\mathrm{G}}$. The convex set $\mathscr{C} = \mathbb{R}_0^{2+}$ is marked in light gray. Note that the classical JOR and SOR schemes converge towards the point $\mathbf{G}^{-1}\mathbf{c}$, i.e. versus the solution of the underlying linear system of equations. However, if $\mathbf{G}^{-1}\mathbf{c} \notin \mathscr{C}$, then the projective JORprox and SORprox schemes do not converge versus this solution due to the additional projections which eliminate the classical JOR or SOR steps. Such situations are indicated by gray/black checkered lines. In Figs. 4.9a,4.10a the loading is $F_y = -0.6$ N and $F_\varphi = 0.2$ Nm, which causes the first unilateral contact to open. Figures 4.9b,4.10b have been calculated using the loading $F_y = -1.2$ N and $F_\varphi = -1$ Nm, for which the second unilateral contact will open. Both unilateral contacts open in Figs. 4.9c,4.10c, and the corresponding loading is $F_y = 0.6$ N and $F_\varphi = 0$ Nm. Finally, in Figs. 4.9d,4.10d the loading is $F_y = -1.2$ N and $F_\varphi = -0.15$ Nm which yields both unilateral contacts to remain closed. Note also the skewed projection to the Λ_2 axis in Fig. 4.10a, which has the same direction as the projection to the Λ_2-axis with respect to $||\cdot||_{\mathrm{G}}$. The projection to the Λ_1-axis in Fig. 4.10b remains orthogonal. In this context it has to be pointed out that systems with more than two unilateral contacts become more complicated.

4.2.7.2 Iterative Solution Process

In this example a planar system which consisting of a block on a surface with external loads F_x, F_y, M is treated [91,93]. The contact between the block and the surface is modeled by two unilateral contacts and two planar friction elements, which act on the block's lower corners (Fig. 4.11). It is assumed that all set-valued force laws can

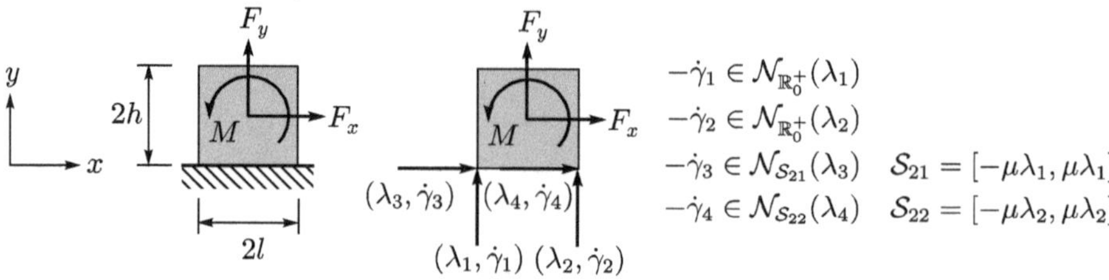

Fig. 4.11 Planar model of a block on a surface

be expressed on acceleration level, i.e. $g_i = 0$ and $\gamma_i = 0$ for the unilateral contacts and $\gamma_i = 0$ for friction elements. For such a case, the equations of motion are

$$\underbrace{\begin{pmatrix} m & 0 & 0 \\ 0 & m & 0 \\ 0 & 0 & J_s \end{pmatrix}}_{\mathbf{M}} \underbrace{\begin{pmatrix} \ddot{x} \\ \ddot{y} \\ \ddot{\varphi} \end{pmatrix}}_{\dot{\mathbf{u}}} - \underbrace{\begin{pmatrix} 0 & 0 & 1 & 1 \\ 1 & 1 & 0 & 0 \\ -\ell & \ell & h & h \end{pmatrix}}_{\mathbf{W}} \underbrace{\begin{pmatrix} \lambda_1 \\ \lambda_2 \\ \lambda_3 \\ \lambda_4 \end{pmatrix}}_{\lambda} - \underbrace{\begin{pmatrix} F_x \\ F_y \\ M \end{pmatrix}}_{\mathbf{h}} = 0. \tag{4.71}$$

The parameters are $m = 2.1$, $J_s = 0.018$, $\ell = 0.15$ and $h = 0.05$, for which the λ_i-equations become

$$\lambda_1 = \text{prox}_{\mathbb{R}_0^+}(\lambda_1 - r_1(\ \ 1.8\lambda_1 - 0.8\lambda_2 - 0.4\lambda_3 - 0.4\lambda_4 + c_1)), \tag{4.72}$$
$$\lambda_2 = \text{prox}_{\mathbb{R}_0^+}(\lambda_2 - r_2(-0.8\lambda_1 + 1.8\lambda_2 + 0.4\lambda_3 + 0.4\lambda_4 + c_2)), \tag{4.73}$$
$$\lambda_3 = \text{prox}_{\mathscr{S}_{21}}(\lambda_3 - r_3(-0.4\lambda_1 + 0.4\lambda_2 + 0.6\lambda_3 + 0.6\lambda_4 + c_3)), \tag{4.74}$$
$$\lambda_4 = \text{prox}_{\mathscr{S}_{22}}(\lambda_4 - r_4(-0.4\lambda_1 + 0.4\lambda_2 + 0.6\lambda_3 + 0.6\lambda_4 + c_4)), \tag{4.75}$$

where c_i contains the terms $\mathbf{c} = \mathbf{W}^\top\mathbf{M}^{-1}\mathbf{h}$. The matrix $\mathbf{G} = \mathbf{W}^\top\mathbf{M}^{-1}\mathbf{W}$ is neither strictly diagonal dominant nor positive definite, but only positive semidefinite. Thus, convergence can not be guaranteed, and the system (4.72-4.75) might have non-unique solutions for the forces λ.

As an example, consider the case that the block remains in the stick state, which can be realized by moderate or vanishing loading (F_x, F_y, M). For this situation one can show that none of the projective equations project on their sets ($\text{prox}_{\mathscr{C}_i}(x_i) \equiv x_i$). As a consequence, (4.72-4.75) reduce to a set of *linear* equations which are not independent due to the semi-definiteness of $\mathbf{G}$, and the forces λ_i can not be determined uniquely. However, it is already satisfactory if the algorithm returns just one of those

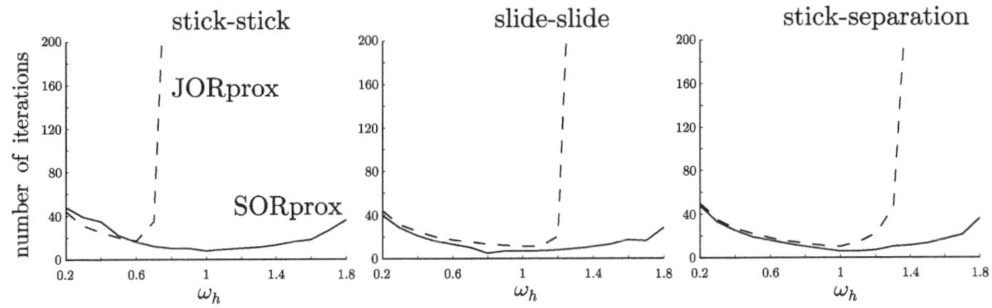

Fig. 4.12 Convergence of the JORprox (dashed) and the SORprox (solid) method. The number of iterations is shown as a function of the relaxation parameter ω_h [91,93]

solutions, because in this example all possible force distributions lead to the same (unique) generalized accelerations $\dot{\mathbf{u}}$. If one of the unilateral contacts is going to open or if the friction elements begin to slide, then the forces $\boldsymbol{\lambda}$ are expected to be unique. In these situations, the solution $\boldsymbol{\lambda}$ of the non-smooth problem (4.72-4.75) is at a point at which some of the $\text{prox}_{\mathscr{C}_i}$-functions really project, which makes the *non*-linear equations of the system (4.72-4.75) independent. Note in this context that the formulation of the $\boldsymbol{\lambda}_i$-equations (4.72-4.75) does not make any assumptions whether the solutions are unique or non-unique, i.e. the system (4.72-4.75) might have unique solutions or not. This would not be the case if the system would be described by $\dot{\gamma}_i$-equations, as these equations request the inverse of $\mathbf{G}$. Using a pseudo-inverse instead of the non-existent inverse induces assumptions on the uniqueness of the solution. Therefore, systems with a non-regular matrix $\mathbf{G}$ can only be treated by $\boldsymbol{\lambda}_i$-equations.

The projective equations (4.72-4.75) are solved by the JORprox and the SORprox method. In figure 4.12 the number of iteration steps is shown in dependence of the relaxation factor ω_h. It is assumed that all set-valued force laws have the same relaxation parameter. Convergence can not be guaranteed, but the iteration converges anyway for some ω_h. In the first plot, both unilateral contacts are closed and both friction elements stick. The solution is non-unique. The plot depicts the number of iteration steps of the JORprox method (dashed) and of the SORprox method (solid). In the second plot, both unilateral contacts are closed and both friction elements slide. The third plot is associated with one closed unilateral contact. The corresponding friction element sticks. The other unilateral contact opens.

4.3 Alternative Approaches

This section presents two alternative approaches which yield the same set of projective $\boldsymbol{\lambda}_i$-equations as the augmented Lagrangian approach.

4.3.1 Uncoupling the Set-Valued Force Laws

This approach rewrites the n local inclusions (4.1) in a convenient way and uses relation (2.17) to derive a set of projective λ_i-equations. Consider the representation of the non-smooth problem by n the local inclusions (4.1). We separate the term $\mathbf{G}_{ii}\lambda_i$ in order to obtain

$$-(\mathbf{G}_{ii}\lambda_i + \sum_{\substack{j=1\\ j\neq i}}^{n} \mathbf{G}_{ij}\lambda_j + \mathbf{c}_i) \in \mathcal{N}_{\mathcal{C}_i}(\lambda_i), \quad i = 1\ldots n, \tag{4.76}$$

which yields

$$-\mathbf{G}_{ii}^{-1}(\sum_{\substack{j=1\\ j\neq i}}^{n} \mathbf{G}_{ij}\lambda_j + \mathbf{c}_i) \in \lambda_i + \mathbf{G}_{ii}^{-1}\mathcal{N}_{\mathcal{C}_i}(\lambda_i), \quad i = 1\ldots n. \tag{4.77}$$

The right-hand side of (4.77) characterizes the behaviour of the i-th set-valued force law. The left-hand side defines a generalized loading κ_i acting on the set-valued force law i. This loading also incorporates by the term $\mathbf{G}_{ii}^{-1}\mathbf{G}_{ij}$ the influence of the j-th set-valued force law on the i-th set-valued force law. Thus, the whole coupling between the set-valued force laws is considered as loading. By writing the local inclusions (4.1) in the form (4.77) it is possible to transform each inclusion separately into projective equations by using the natural map (2.17), i.e.

$$\lambda_i + \mathbf{R}_i^{-1}\mathcal{N}_{\mathcal{C}_i}(\lambda_i) \ni \kappa_i \quad \Leftrightarrow \quad \lambda_i = \mathrm{prox}_{\mathcal{C}_i}^{\mathrm{R}_i}(\kappa_i). \tag{4.78}$$

A more general approach adds the term $\mathbf{R}\lambda$ on both sides of the global inclusion (4.2),

$$\mathbf{R}\lambda - (\mathbf{G}\lambda + \mathbf{c}) \in \mathbf{R}\lambda + \mathcal{N}_{\mathcal{C}}(\lambda). \tag{4.79}$$

Multiplication by $\mathbf{R}^{-1}$ yields

$$\lambda - \mathbf{R}^{-1}(\mathbf{G}\lambda + \mathbf{c}) \in \lambda + \mathbf{R}^{-1}\mathcal{N}_{\mathcal{C}}(\lambda), \tag{4.80}$$

from which the λ-equation (4.23) can be obtained using relation (2.17). Thus the inclusion (4.2) is brought into the special form (4.79) which is suitable for (2.17) and is transformed straightforward into a global projective λ-equation. Of course the global λ-equation can be split into the local λ_i-equations if $\mathbf{R}$ is chosen according to (2.30).

4.3.2 Exact Regularization

In the following, a third way to obtain the set of λ_i-equations is discussed, but now based on the concept of exact regularization [57]. The main difficulty when dealing with set-valued force laws is the fact that a whole set is obtained for some arguments. Take for instance the set-valued force law of the unilateral contact, i.e. $\lambda_i \in \partial\Psi^\star_{\mathbb{R}_0^+}(-g_i)$. While λ_i is uniquely determined for $g_i > 0$, i.e. $\lambda_i = 0$ for $g_i > 0$, the case of g_i being zero yields only $\lambda_i \in \mathbb{R}_0^+$. One way to avoid this problem is to regularize the set-valued force laws. For example, the set-valued part of the unilateral contact law $\lambda_i \in \partial\Psi^\star_{\mathbb{R}_0^+}(-g_i)$ can be regularized as a straight line with slope r and intersection $\lambda_i^\star$ at the ordinate. This leads to a regularized force law

$$\lambda_i = f_{\text{reg}}(g_i) = \text{prox}_{\mathbb{R}_0^+}(\lambda_i^\star - r\,g_i). \tag{4.81}$$

The function $f_{\text{reg}}(g_i)$ has a unique value λ_i for every function argument g_i. Of course, the regularized and the set-valued force law are different and do not describe the same behaviour. In Fig. 4.13, the set-valued force law $\lambda_i \in \partial\Psi^\star_{\mathbb{R}_0^+}(-g_i)$ is plotted in light gray, the regularized force law $\lambda_i = f_{\text{reg}}(g_i)$ law in dashed dark gray. The set-valued force law and the regularized force law coincide if

$$(g_i \geq 0\ ,\ \lambda_i = \lambda_i^\star = 0), \tag{4.82}$$
$$(g_i = 0\ ,\ \lambda_i = \lambda_i^\star \geq 0). \tag{4.83}$$

Thus, every point (g_i, λ_i) that satisfies

$$\lambda_i = \lambda_i^\star = f_{\text{reg}}(g_i) = \text{prox}_{\mathbb{R}_0^+}(\lambda_i^\star - r\,g_i) \tag{4.84}$$

is a point of the set-valued function $\lambda_i \in \partial\Psi^\star_{\mathbb{R}_0^+}(-g_i)$. The idea of exact regularization is to shift the regularized function to the point at which the true solution, i.e. the solution of the set-valued function would be. Considering the unilateral contact, the regularized law is equivalent to modeling the unilateral contact by a spring, where r is the spring stiffness and $\lambda_i^\star$ the force at $g_i = 0$. Exact regularization adjusts iteratively the spring parameter $\lambda^\star$ in such way that a positive force $\lambda_i > 0$ yields

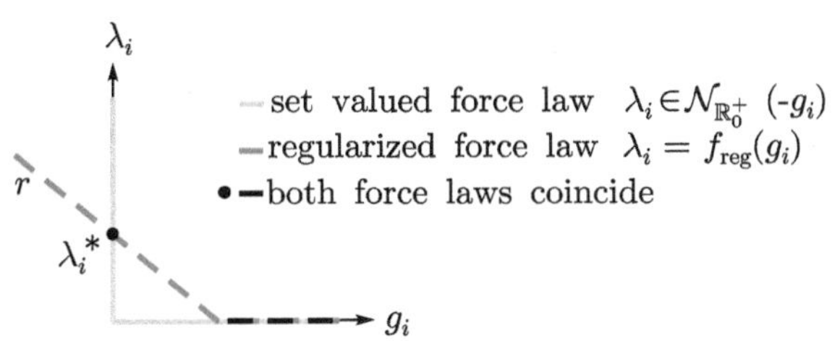

Fig. 4.13 Regularization of $\lambda_i \in \partial\Psi^\star_{\mathbb{R}_0^+}(-g_i) \equiv \mathcal{N}_{\mathbb{R}_0^+}(-g_i)$

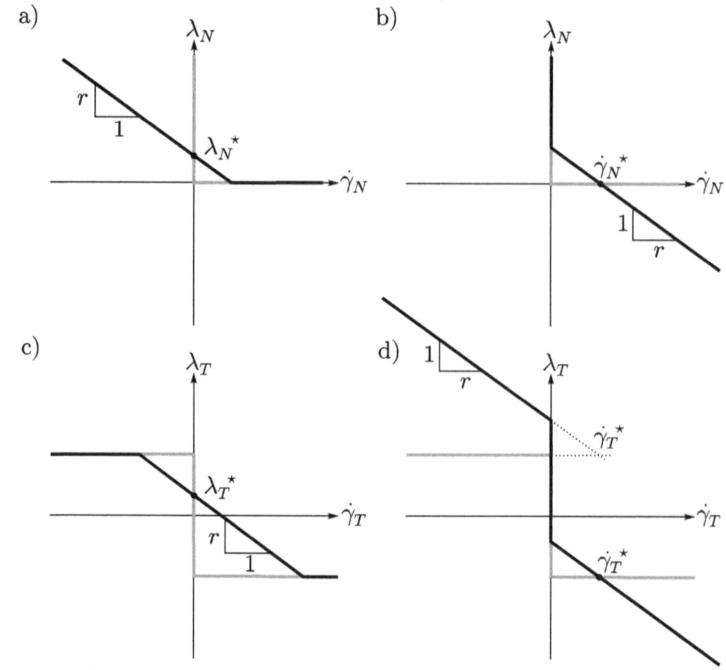

inclusion	projective equation	figure
$\lambda_N \in \partial\Psi^\star_{\mathbb{R}_0^+}(-\dot\gamma_N)$	$\lambda_N = \lambda_N^\star = \ \mathrm{prox}_{\mathbb{R}_0^+}(\lambda_N^\star - r\,\dot\gamma_N)$	*a)*
$-\dot\gamma_N \in \mathcal{N}_{\mathbb{R}_0^+}(\lambda_N)$	$\dot\gamma_N = \dot\gamma_N^\star = -\frac{1}{r}\mathrm{vdst}_{\mathbb{R}_0^+}(-r\,\dot\gamma_N^\star + \lambda_N)$	*b)*
$\lambda_T \in \partial\Psi^\star_{\mathcal{S}_p}(-\dot\gamma_T)$	$\lambda_T = \lambda_T^\star = \ \mathrm{prox}_{\mathcal{S}_p}(\lambda_T^\star - r\,\dot\gamma_T)$	*c)*
$-\dot\gamma_T \in \mathcal{N}_{\mathcal{S}_p}(\lambda_T)$	$\dot\gamma_N = \dot\gamma_N^\star = -\frac{1}{r}\mathrm{vdst}_{\mathcal{S}_p}(-r\,\dot\gamma_N^\star + \lambda_T)$	*d)*

Fig. 4.14 Nonlinear equations describing the set-valued force laws for unilateral contact and planar friction with $\mathscr{S}_p = [-a, a]$. The regularized force law is shifted to a position such that it intersects the set-valued branch of the inclusion in the point at which the solution is [91]

$g_i = 0$. Figure 4.14 summarizes the set-valued force laws for unilateral contact and planar friction on acceleration level together with their associated exact regularization. Elimination of the relative accelerations $\dot\gamma_i$ or of the forces λ_i using (4.1) yields finally the λ_i and the $\dot\gamma_i$-equations, respectively. The iterative solution of the λ_i-equations can then be interpreted in the following way: The non-smooth problem is solved for an initial regularization $\lambda_0^\star$. Based on the resulting forces λ, a better regularization is obtained. This procedure is repeated until a regularization is found for which regularized and set-valued force laws coincide.

4.4 Summary

In this chapter it has been discussed how inclusion problems can be solved. These problems can either be stated by n local or one global inclusion (4.1) and (4.2), or the problem can be defined by an associated constrained optimization problem (4.4).

The augmented Lagrangian approach has been applied to the optimization problem to obtain a global λ or $\dot{\gamma}$-equation, which are projective equations with respect to the norm $||\cdot||_{\mathrm{R}}$. Choosing $\mathbf{R} = \mathbf{G}$ yields the solution of the non-smooth problem within one projection. Unfortunately, this projection is so far as expensive as a combinatoric solution of the non-smooth problem would be. Choosing the matrix $\mathbf{R}$ as block matrix according to (2.30) allows to split the global λ or $\dot{\gamma}$-equation into n local λ_i or $\dot{\gamma}_i$-equations, which must be solved iteratively. Only the λ_i-equations have been examined further on, because a non-regular matrix $\mathbf{G}$ makes the formulation of $\dot{\gamma}$-equations impossible. When choosing $\mathbf{R}_i = \mathbf{G}_{ii}$, the set-valued force law i can be solved straightforward if all other forces $\lambda_{i \neq j}$ are known. However, the possible projection within the λ_i-equations is with respect to $||\cdot||_{\mathrm{G}_{ii}}$, which might be cumbersome. Choosing the matrix $\mathbf{R}_{\mathrm{I}} = \mathbf{R}_i = \frac{1}{r_i}\mathbf{I}$ with $r_i > 0$ allows for using the Euclidian norm for the evaluation of the $\mathrm{prox}_{\mathscr{C}_i}^{\mathrm{R}}$ function. Based on this choice of $\mathbf{R}$, two iterative schemes are suggested to solve the λ_i-equations, the JORprox and the SORprox scheme. The JORprox scheme consist of a Jacobi relaxation step for the linear system $\mathbf{G}\lambda + \mathbf{c} = 0$ followed by a possible projection that ensures that the force λ_i is within the set of admissible forces $\mathscr{C}_i$. Convergence of the JORprox method has been proven for strictly diagonal dominant matrices $\mathbf{G}$ in the absence of Coulomb friction. Other than the JORprox method, the SORprox method makes use of the fact that when calculating $\lambda_i^{\nu+1}$ already updated forces $\lambda_{j<i}^{\nu+1}$ are available. The SORprox scheme solves the underlying linear system $\mathbf{G}\lambda + \mathbf{c} = 0$ by a Gauss Seidel relaxation method, and is expected to converge for positive definite matrices $\mathbf{G}$. However, a proof of this claim is missing. The SORprox scheme has shown good convergence results in practical applications.

In addition, it has been examined how the inclusion associated with the i-th set-valued force law can be separated in terms which describe the behaviour of the set-valued force law and in terms which define the influence of the set-valued force law $j \neq i$ on the set-valued force law i. The latter terms can be considered as generalized loading on the i-th set-valued force law, which allows for solving the set-valued force law i straightforward. A more general approach adds on both sides of the global inclusion (4.2) the term $\mathbf{R}\lambda$ and transform this inclusion into the λ-equation using relation (2.17). Doing so, the corresponding optimization problem is modified to become equivalent to the definition of the $\mathrm{prox}_{\mathscr{C}}^{\mathrm{R}}$ function.

A third method is to regularize the set-valued force laws and to shift them into a position at which they coincide with the original ones. This idea is known as exact regularization. Exact regularization links the two main approaches in non-smooth mechanics; the regularized and the set-valued approach.

Note the different interpretations of the matrix $\mathbf{R}$ and its possible scalar equivalent r_i. In the augmented Lagrangian approach they appear as penalty terms. On the other hand, $\mathbf{R}$ defines the norm $||\cdot||_{\mathrm{R}}$ in which the $\mathrm{prox}_{\mathscr{C}}^{\mathrm{R}}$ function is evaluated. Choosing $\mathbf{R} = \mathbf{G}$ would solve the non-smooth problem straightforward within one projection. However, this projection is so expensive that a diagonal matrix $\mathbf{R} = \mathbf{R}_{\mathrm{I}}$ is chosen such that the norm $||\cdot||_{\mathrm{R_I}}$ comes closest to $||\cdot||_{\mathrm{G}}$. Note that the choice of a diagonal $\mathbf{R} = \mathbf{R}_{\mathrm{I}}$ allows for using the Euclidian norm for the projection. Furthermore, the scalars r_i define the influence between the set-valued force laws. If r_i is chosen small, then the influence of the other set-valued force laws $j \neq i$ on the set-valued force law i decreases. In the framework of exact regularization, the scalars r_i represent the slope of the regularized laws. Considering the iterative solution procedure by the JORprox or SORprox methods, r_i is associated with the relaxation factor ω_h.

Chapter 5
Time-Stepping

In this chapter time-stepping schemes are investigated more closely. In Sect. 5.1 possible discretizations for differential algebraic systems are discussed. These discretizations build the fundament of the time-stepping schemes. Special attention is paid to the drift phenomenon of acceleration and velocity based constraint formulations. We discuss the discretization of index-1, index-2 and index-3 DAE's. The latter are treated either by a GGL [86] or a preconditioning approach [19]. Furthermore, the nonlinear gap function of an index-3 problem is linearized which yields an index-2-like scheme with additional explicit drift stabilization. In Sect. 5.2, the behaviour of non-smooth systems is discussed. The time evolution of such systems is split into different smooth parts, which are described by an underlying DAE. Event-driven methods do not discretize the non-smooth system itself but the different smooth parts, and detect switching points at which the underlying DAE and thus the discretization changes. Time-stepping schemes directly discretize the non-smooth system regardless of switching points, see Sect. 5.3. Moreau's time-stepping method is discussed in detail. This prominent method will also serve as base method for chapter 6. In addition, a brief review on existing time-stepping methods is given. First, the methods of Jean [47] and Paoli/Schatzman [68] are discussed. Furthermore, the contributions of Stiegelmeyr [85], Funk [36], Foerg [33] and Pfeiffer [72], as well as the methods of Stewart/Trinkle [84], Anitescu/Potra [10, 11], Anitescu/Hart [9], Potra et al [74] and Gavrea et al [37] are reviewed. In addition, some ideas for two new time-stepping schemes which incorporate the GGL or the preconditioning approach are given. A discussion on the classification of the different schemes is found in Sect. 5.3.9.

5.1 Discretization of Differential Algebraic Equations

Differential algebraic equations have been used in Sect. 3.1 to describe mechanical systems subjected to constraints. In this section, the numerical treatment of such DAE's is discussed. The aim of this section is to prepare the reader for the discussion of time-stepping schemes, which are very closely related to the dis-

C. Studer: Numerics of Unilateral Contacts and Friction, LNACM 47, pp. 59–97.
springerlink.com

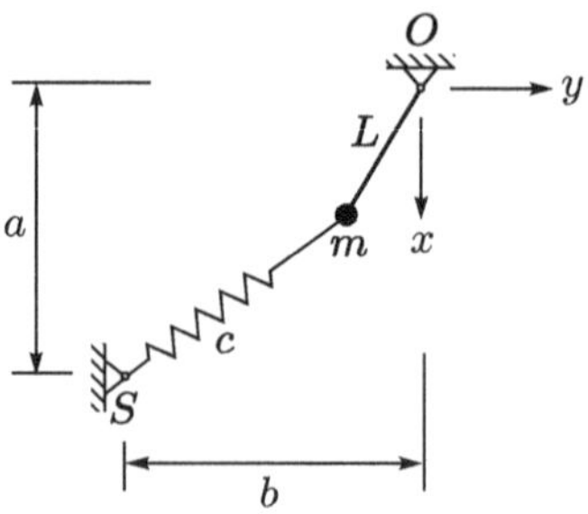

Fig. 5.1 Pendulum system

cretization of DAE's. A differential algebraic equation of index-1 can be discretized and solved in a straightforward way. The handling of an index-2 or an index-3 DAE is a more challenging task. In the following, the discretization of a differential algebraic equation is explained by an example. This example intends to give the reader a brief introduction to the topic. The reader is referred to the literature on DAE's, see for example [26, 46, 58, 86]. Consider a mathematical pendulum with mass m and length L, which has an additional spring with stiffness c attached at its end, see Fig. 5.1. The spring is unstressed for $x = a$ and $y = -b$. The position of the spring's support is denoted by $\mathbf{r}_{OS} = (a\ -b)^\top$, where the hinge of the pendulum defines the origin of the coordinate system. The position and velocity of the point mass is described by $\mathbf{q} = (x\ y)^\top$ and $\mathbf{u} = \dot{\mathbf{q}}$, respectively. The associated DAE yields

$$0 = m\mathbf{I}\,\dot{\mathbf{u}} - (\mathbf{r}_{OS} - \mathbf{q})\,c - \frac{\mathbf{q}}{||\mathbf{q}||}\lambda := \mathbf{M}\dot{\mathbf{u}} - \mathbf{h} - \mathbf{w}\lambda, \tag{5.1}$$

$$0 = \dot{\mathbf{q}} - \mathbf{u}, \tag{5.2}$$

$$0 = \sqrt{\mathbf{q}^\top\mathbf{q}} - L := g, \tag{5.3}$$

which is a DAE of index-3. It is distinguished between the equations of motion (5.1-5.2) and the algebraic gap function (5.3), which defines the constraint. The discrete form of (5.2) is referred to as position update formula. The DAE (5.1-5.3) can be solved in various ways. First, the index-3 problem is transformed into an index-2 or an index-1 problem by differentiating the gap function (5.3) with respect to the time, i.e.

$$0 = \dot{g} = \frac{\mathbf{q}^\top}{||\mathbf{q}||}\mathbf{u} = \mathbf{w}^\top\mathbf{u}, \tag{5.4}$$

$$0 = \ddot{g} = \frac{\mathbf{q}^\top}{||\mathbf{q}||}\dot{\mathbf{u}} + \frac{\mathbf{u}^\top\mathbf{u}}{||\mathbf{q}||} - \frac{(\mathbf{q}^\top\mathbf{u})^2}{||\mathbf{q}||^3} = \mathbf{w}^\top\dot{\mathbf{u}} + \hat{\zeta}. \tag{5.5}$$

The resulting index-2 and index-1 problems can be discretized in a straightforward way. Methods which solve directly the index-3 problem are for example the Baumgarte stabilization [86], the GGL method [86] and a preconditioning method [19].

Furthermore, a linearization of the gap function (5.3) can be done, which yields an index-2-like scheme with an additional constraint stabilization.

Consider a time step with step size Δt. The states $\mathbf{q}$ and $\mathbf{u}$ at the begin and end point are indicated by the indices B and E, respectively. Furthermore, $\mathbf{h}_M = \mathbf{h}(\mathbf{q}_M)$, $\mathbf{w}_M = \mathbf{w}(\mathbf{q}_M)$ and $\hat{\zeta}_M = \hat{\zeta}(\mathbf{q}_M, \mathbf{u}_B)$ are defined at the midpoint $\mathbf{q}_M = \mathbf{q}_B + \mathbf{u}_B \frac{\Delta t}{2}$. The following parameters have been used in the numerical simulations: $\Delta t = 0.05$s, $m = 1$ kg, $L = 1$ m, $a = b = 2$ m and $c = 10$N/m.

5.1.1 Index-1, Acceleration Level

Replacing condition (5.3) by (5.5) yields an index-1 DAE with a possible discretization

$$0 = \mathbf{M}\frac{\mathbf{u}_E - \mathbf{u}_B}{\Delta t} - \mathbf{h}_M - \mathbf{w}_M \lambda, \tag{5.6}$$

$$0 = \frac{\mathbf{q}_E - \mathbf{q}_B}{\Delta t} - \frac{\mathbf{u}_E + \mathbf{u}_B}{2}, \tag{5.7}$$

$$0 = \mathbf{w}_M^{\top}\frac{\mathbf{u}_E - \mathbf{u}_B}{\Delta t} + \hat{\zeta}_M. \tag{5.8}$$

This linear system of equations (5.6-5.8) can be solved for the unknowns $\mathbf{q}_E$,$\mathbf{u}_E$ and λ to obtain

$$\lambda = -\left(\mathbf{w}_M^{\top}\mathbf{M}^{-1}\mathbf{w}_M\right)^{-1}\left(\mathbf{w}_M^{\top}\mathbf{M}^{-1}\mathbf{h}_M + \hat{\zeta}_M\right), \tag{5.9}$$

$$\mathbf{u}_E = \mathbf{u}_B + \Delta t\, \mathbf{M}^{-1}\left(\mathbf{h}_M + \mathbf{w}_M \lambda\right), \tag{5.10}$$

$$\mathbf{q}_E = \mathbf{q}_B + \Delta t\, \frac{\mathbf{u}_E + \mathbf{u}_B}{2}. \tag{5.11}$$

Using condition (5.5) instead of (5.3) causes drift. The condition $g = 0$ and $\ddot{g} = 0$ are equivalent if at an arbitrary time $t = t_0$ the conditions $g(\mathbf{q}(t_0)) = 0$ and $\gamma(\mathbf{q}(t_0), \mathbf{u}(t_0)) = 0$ are fulfilled, as already discussed in Sect. 3.1. We place ourselves at a time $t = t_0$ on the manifolds $g = 0$ and $\gamma = 0$ by requesting $g(\mathbf{q}(t_0)) = 0$ and $\gamma(\mathbf{q}(t_0), \mathbf{u}(t_0)) = 0$, and we use the condition $\dot{\gamma} = 0$ to stay on the manifold for $t > t_0$. When using the discrete scheme (5.6-5.8), only an approximation of the derivative $\dot{\gamma}$ is available, and the solution of the problem might therefore drift away from the manifolds $g = 0$ and $\gamma = 0$. The results of a simulation which uses the discretization (5.6-5.8) are depicted in Fig. 5.2. Figure 5.2a visualizes the path in the x-y-plane. The drift is obvious. Figures 5.2b and 5.2c depict the phase plots x versus $\dot{x}$ and y versus $\dot{y}$. Note that a horizontal shift corresponds to the drift in the positions, a vertical shift is associated with a drift in the velocities. As it can been seen, the presented solution is not acceptable and the step-size should be reduced to prevent such pronounced drift phenomenons.

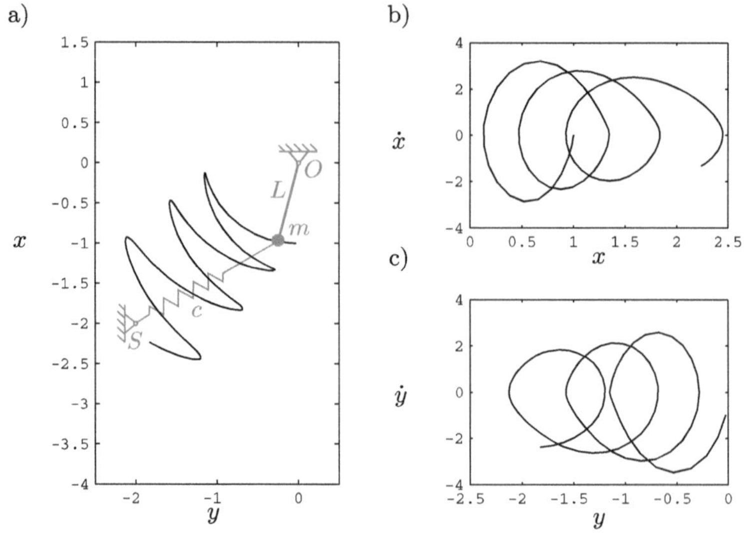

Fig. 5.2 Index-1 discretization. Figure **a** visualizes the path of the system in the x-y-plane. Figures **b** and **c** depict the phase plots x versus $\dot{x}$ and y versus $\dot{y}$

5.1.2 *Index-2, Velocity Level*

Instead of (5.3) the condition (5.4) is used to obtain an index-2 DAE with a possible discretization

$$0 = \mathbf{M}\,\frac{\mathbf{u}_E - \mathbf{u}_B}{\Delta t} - \mathbf{h}_M - \mathbf{w}_M \lambda, \tag{5.12}$$

$$0 = \frac{\mathbf{q}_E - \mathbf{q}_B}{\Delta t} - \frac{\mathbf{u}_E + \mathbf{u}_B}{2}, \tag{5.13}$$

$$0 = \mathbf{w}_M^{\top}(\mathbf{u}_E + \varepsilon \mathbf{u}_B). \tag{5.14}$$

Note the coefficient ε in (5.14) which allows for modifying the discretized constraint. This coefficient will later be identified as restitution coefficient within Moreau's time stepping scheme, see Sect. 5.3.2. Again, one can solve for the unknowns $\mathbf{q}_E$,$\mathbf{u}_E$ and λ, which yields

$$\lambda\,\Delta t = -\left(\mathbf{w}_M^{\top}\mathbf{M}^{-1}\mathbf{w}_M\right)^{-1}\left(\mathbf{w}_M^{\top}\mathbf{M}^{-1}\mathbf{h}_M\Delta t + (1+\varepsilon)\mathbf{w}_M^{\top}\mathbf{u}_B\right), \tag{5.15}$$

$$\mathbf{u}_E = \mathbf{u}_B + \mathbf{M}^{-1}\left(\mathbf{h}_M\Delta t + \mathbf{w}_M\lambda\,\Delta t\right), \tag{5.16}$$

$$\mathbf{q}_E = \mathbf{q}_B + \Delta t\,\frac{\mathbf{u}_E + \mathbf{u}_B}{2}. \tag{5.17}$$

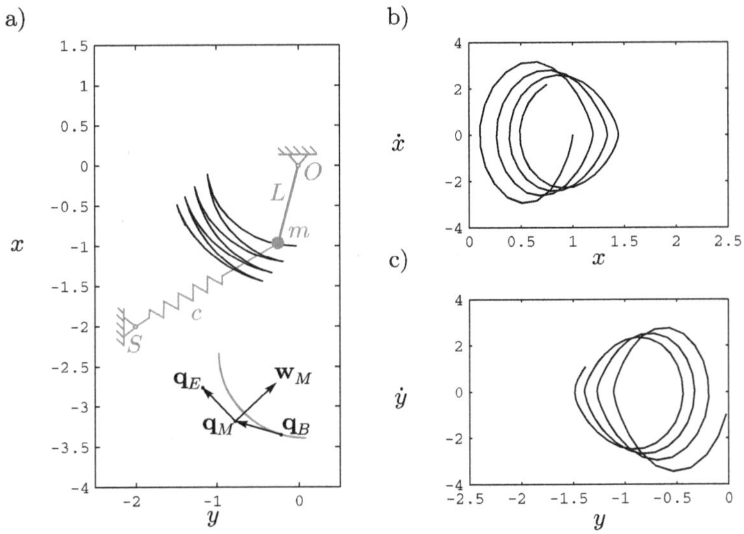

Fig. 5.3 Index-2 discretization with $\varepsilon = 0$. Figure **a** visualizes the path of the system in the x-y-plane. Due $\varepsilon = 0$ the solution moves rapidly away from the manifold $g = 0$. Figures **b** and **c** depict the phase plots x versus $\dot{x}$ and y versus $\dot{y}$

Note that the calculation of λ becomes problematic if $\Delta t \to 0$. This holds also in general, i.e. the linear system of equations associated with an index-2 discretization is ill-conditioned due to λ in case of $\Delta t \to 0$. Therefore, the sketched approach to solve an index-2 DAE is invalid. The problem can be fixed by defining a discrete percussion $\hat{P} = \lambda\,\Delta t$, which replaces λ as unknown. The representation of an index-3 problem by an index-2 problem is subjected to drift in the positions. Figures 5.3 and 5.4 show the drift which occurs when using the discretization (5.12-5.14) with $\varepsilon = 0$ and $\varepsilon = 1$. Using $\varepsilon = 0$ yields the results depicted in Fig. 5.3. The positions are subjected to drift, note especially the horizontal shift in the phase plots. The velocities and thus the kinetic energy decrease because of the fully implicit position update formula. Choosing $\varepsilon = 1$ reduces the drift, see Fig. 5.4. In this case, the constraint (5.14) becomes a reflection law. Note that these observations can not be generalized to arbitrary systems. In case of circle-like trajectories, a reflection law might be preferable. Note that the choice $\varepsilon = 1$ might lead to oscillations. Only the condition $\mathbf{w}^\top(\mathbf{u}_E + \mathbf{u}_B) = \gamma_E + \gamma_B = 0$ has to be fulfilled in this case. This condition does not prevent oscillations in γ, i.e. if we have $\gamma_B \neq 0$ due to an numerical error, then all further γ_B and γ_E will also be unequal to zero. In order to prevent these oscillations, ε can be chosen close to one, which allows for a weak damping in the constraint.

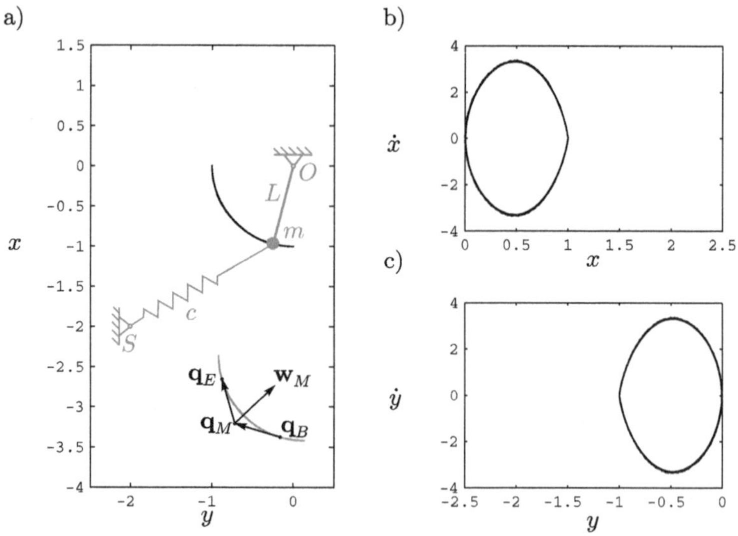

Fig. 5.4 Index-2 discretization with $\varepsilon = 1$. Figure **a** visualizes the path of the system in the x-y-plane. Due $\varepsilon = 1$, the constraint (5.14) is a reflection law and the drift is minimal. Figures **b** and **c** depict the phase plots x versus $\dot{x}$ and y versus $\dot{y}$

Remark

Note the link between the index-2 discretization and the equality of measures. Instead of the differential equation (5.1) the associated equation of measures $\mathbf{M}\,d\mathbf{u} - \mathbf{h}\,dt - \mathbf{w}\,dP = 0$ is discretized to obtain $\mathbf{M}\,(\mathbf{u}_E - \mathbf{u}_B) - \mathbf{h}_M \Delta t - \mathbf{w}_M \hat{P} = 0$. Conditions on velocity level should not be enforced by finite forces λ.

5.1.3 Index-3, Displacement Level

A possible discretization of the index-3 DAE (5.1-5.3) is

$$0 = \mathbf{M}\,\frac{\mathbf{u}_E - \mathbf{u}_B}{\Delta t} - \mathbf{h}_M - \mathbf{w}_M \lambda, \tag{5.18}$$

$$0 = \frac{\mathbf{q}_E - \mathbf{q}_B}{\Delta t} - \frac{\mathbf{u}_E + \mathbf{u}_B}{2}, \tag{5.19}$$

$$0 = \sqrt{\mathbf{q}_E^{\top}\,\mathbf{q}_E} - L. \tag{5.20}$$

Solving the system (5.18-5.20) for the unknowns $\mathbf{q}_E$,$\mathbf{u}_E$ and λ becomes more complicated than in the index-1 or index-2 case, as the constraint (5.20) is no longer a linear equation. The system (5.18-5.20) is solved as follows: By (5.18) the velocity

$\mathbf{u}_E$ can be written as a function of $\lambda\,\Delta t$. This relation can be inserted into (5.19), which yields $\mathbf{q}_E$ as function of $\lambda\,\Delta t^2$. Replacing $\mathbf{q}_E$ in (5.20) by this function yields a nonlinear equation, which can be solved for λ. In case of $\Delta t \to 0$, the calculation of both the force λ and the velocity $\mathbf{u}_E$ becomes a problem. In the following, some approaches to solve the problem are outlined.

5.1.3.1 Baumgarte Stabilization

The idea behind the Baumgarte stabilization is to formulate the index-3 DAE as an index-1 DAE and to use a spring-damper element to control the drift, see for example [46, 86]. The constraint equation $g = 0$ is replaced by $\ddot{g} + 2\alpha\dot{g} + \alpha^2 g = 0$ with $\alpha > 0$. All unknowns are defined in the case $\Delta t \to 0$, and the solution $\mathbf{q}(t)$ and $\mathbf{u}(t)$ is expected to converge towards the manifold $g = 0$ and $\dot{g} = 0$, respectively. However, the choice of a suitable α might be difficult.

5.1.3.2 GGL

The GGL approach was introduced by Gear, Gupta and Leimkuhler, see for example [46, 86]. The approach reformulates the index-3 Problem (5.1-5.3) to arrive at

$$0 = \mathbf{M}\dot{\mathbf{u}} - \mathbf{h} - \mathbf{w}\lambda_{\mathrm{u}}, \tag{5.21}$$

$$0 = \dot{\mathbf{q}} - \mathbf{u} - \mathbf{M}^{-1}\mathbf{w}\lambda_{\mathrm{q}}, \tag{5.22}$$

$$0 = \mathbf{w}^{\top}\mathbf{u}, \tag{5.23}$$

$$0 = \sqrt{\mathbf{q}^{\top}\mathbf{q}} - L, \tag{5.24}$$

The Lagrange multiplier λ_{u} enforces the constraint (5.23) on velocity level, while a second Lagrange multiplier λ_{q} is responsible for the constraint (5.24) on displacement level. Note that the exact solution of (5.21-5.24) yields an index-2 DAE, hence $\lambda_{\mathrm{q}} = 0$. This can be verified by multiplying (5.22) by $\mathbf{w}^{\top}$, i.e. $\mathbf{w}^{\top}\dot{\mathbf{q}} - \mathbf{w}^{\top}\mathbf{u} - \mathbf{w}^{\top}\mathbf{M}^{-1}\mathbf{w}\lambda_{\mathrm{q}} = 0$. The first term is associated with the time derivative of (5.24), i.e. $\mathbf{w}^{\top}\dot{\mathbf{q}} = 0$. The second term vanishes due to (5.23). Thus $\mathbf{w}^{\top}\mathbf{M}^{-1}\mathbf{w}\lambda_{\mathrm{q}} = 0$ which yields $\lambda_{\mathrm{q}} = 0$ for $\mathbf{w}^{\top}\mathbf{M}^{-1}\mathbf{w}$ being regular. The DAE (5.21-5.24) can be discretized in the same manner as an index-2 DAE, i.e.

$$0 = \mathbf{M}(\mathbf{u}_E - \mathbf{u}_B) - \mathbf{h}_M \Delta t - \mathbf{w}_M \hat{P}_{\mathrm{u}}, \tag{5.25}$$

$$0 = (\mathbf{q}_E - \mathbf{q}_B) - \frac{\Delta t}{2}(\mathbf{u}_E + \mathbf{u}_B) - \mathbf{M}_M^{-1}\mathbf{w}_M \hat{P}_{\mathrm{q}}, \tag{5.26}$$

$$0 = \mathbf{w}_M^{\top}(\mathbf{u}_E + \varepsilon\mathbf{u}_B), \tag{5.27}$$

$$0 = \sqrt{\mathbf{q}_E^{\top}\mathbf{q}_E} - L. \tag{5.28}$$

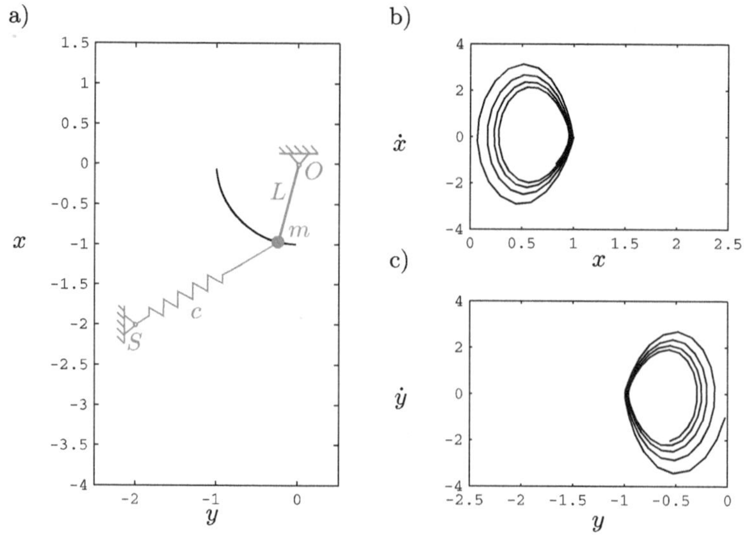

Fig. 5.5 GGL formulation. Figure **a** visualizes the path of the system in the x-y-plane, where $\varepsilon = 0$. Figures **b** and **c** depict the phase plots x versus $\dot{x}$ and y versus $\dot{y}$

Note that λ_{q} and λ_{u} are replaced by the discrete percussions $\hat{P}_{\mathrm{q}} = \lambda_{\mathrm{q}}\Delta t$ and $\hat{P}_{\mathrm{u}} = \lambda_{\mathrm{u}}\Delta t$ to avoid ill-conditioning. The discrete percussion $\hat{P}_{\mathrm{u}}$ enforces the constraint on velocity level, i.e. $\hat{P}_{\mathrm{u}}$ is identical with the Lagrange multiplier within the index-2 solution. Considering our example, the index-2 solution with $\varepsilon = 1$ is not subjected to pronounced drift, see Fig. 5.4, and the GGL formulation becomes superfluous in this case. Therefore, in the following ε is chosen equal to zero in order to underline the action of the GGL formulation. The drift is then prevented by $\hat{P}_{\mathrm{q}}$, which performs a projection to guarantee that the constraint is also fulfilled on displacement level. This projection might increase or decrease the energy of the mechanical system.

The system (5.25-5.28) can be solved for the unknowns $\mathbf{q}_E$, $\mathbf{u}_E$, $\hat{P}_{\mathrm{u}}$ and $\hat{P}_{\mathrm{q}}$. The discrete percussion $\hat{P}_{\mathrm{u}}$ and the velocity $\mathbf{u}_E$ are determined by (5.25) and (5.27), while $\hat{P}_{\mathrm{q}}$ and $\mathbf{q}_E$ follow from (5.26) and (5.28). Note that the latter case requests solving a nonlinear equation. The results of a simulation which uses the discretization (5.25-5.28) with $\varepsilon = 0$ are depicted in Fig. 5.5. Note that the solution is not subjected to drift problems. Consider the phase plots in Figs. 5.5b and 5.5c. Unlike the phase plots in Fig. 5.3, no horizontal shift is present which means that the phase curves of the index-2 solution shown in Fig. 5.3 are shifted together by the projection. Similar to the index-2 solution, the velocities decrease.

5.1.3.3 Preconditioning

Another approach to handle DAE's of index three is presented by Botasso in [19]. Regarding (5.18-5.20), we define $\hat{\mathscr{P}} = \hat{P}\Delta t = \lambda\,\Delta t^2$ and $\hat{\mathbf{u}} = \mathbf{u}\Delta t$. When using these

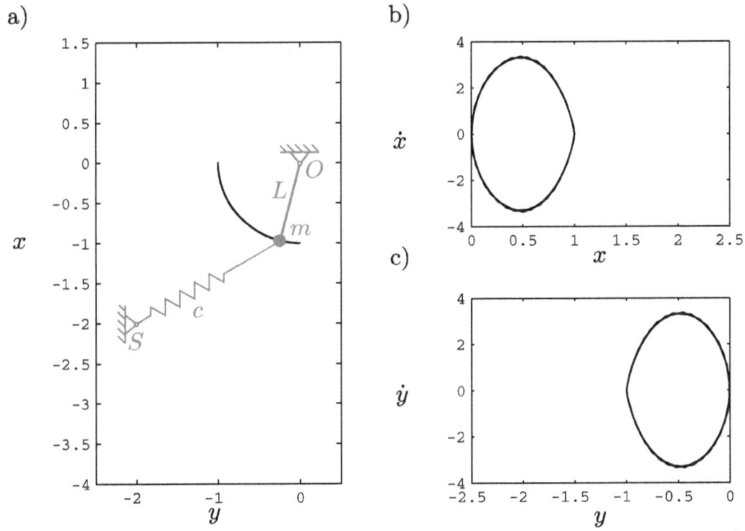

Fig. 5.6 Preconditioning of an index-3 discretization. Figure **a** visualizes the path of the system in the x-y-plane, the Figs. **b** and **c** depict the phase plots x versus $\dot{x}$ and y versus $\dot{y}$

two variables instead of λ and $\mathbf{u}$, then the solution of the index-3 problem is not ill-conditioned anymore. The discretization (5.18-5.20) becomes

$$0 = \mathbf{M}\left(\hat{\mathbf{u}}_E - \hat{\mathbf{u}}_B\right) - \mathbf{h}_M \Delta t^2 - \mathbf{w}_M \hat{\mathscr{P}}, \tag{5.29}$$

$$0 = \mathbf{q}_E - \mathbf{q}_B - \frac{\hat{\mathbf{u}}_E + \hat{\mathbf{u}}_B}{2}, \tag{5.30}$$

$$0 = \sqrt{\mathbf{q}_E^\top \mathbf{q}_E} - L, \tag{5.31}$$

which can be solved for the unknowns $\mathbf{q}_E, \hat{\mathbf{u}}_E$ and $\hat{\mathscr{P}}$ also in the case $\Delta t \to 0$. Note again that a nonlinear equation has to be solved due to (5.31). Figure 5.6 depicts the results of a simulation which uses the discussed preconditioning approach. As for the GGL approach, no drift is present. In addition, the total energy remains constant.

Remarks

Botasso [19] discusses this approach from the viewpoint of preconditioning. It might be also possible to see the approach in the context of equalities of measures. Note that the index-1 problem has been solved by discretizing the differential equation $\mathbf{M}\dot{\mathbf{u}} - \mathbf{h} - \mathbf{W}\lambda$. In order to solve the index-2 problem the equality of measures $\mathbf{M}\,\mathrm{d}\mathbf{u} - \mathbf{h}\,\mathrm{d}t - \mathbf{W}\,\mathrm{d}\mathbf{P}$ has been used. The discussed index-3 approach discretizes $\mathbf{M}\,\mathrm{d}\mathbf{u}\,\mathrm{d}t - \mathbf{h}\,\mathrm{d}t^2 - \mathbf{W}\,\mathrm{d}\mathbf{P}\,\mathrm{d}t$ which might also be written as $\mathbf{M}\,\mathrm{d}^2\mathbf{q} - \mathbf{h}\,\mathrm{d}t^2 - \mathbf{W}\,\mathrm{d}\mathscr{P}$.

Note that $d^2\mathbf{q} = dd\mathbf{q} = d\frac{d\mathbf{q}}{dt}dt = d\mathbf{u}\,dt$. The term $d\mathscr{P} = d\mathbf{P}\,dt$ would correspond to a Lagrangian multiplier which can enforce displacement jumps. Note that such jumps would request infinite impulsive forces and velocities, a fact that corresponds to the ill-conditioning of an index-3 discretization when using only discrete percussions $\hat{P}$ and velocities $\mathbf{u}$.

In the following, we relate the approach to a finite difference discretization $\ddot{\mathbf{q}} = \frac{1}{\Delta t^2}(\mathbf{q}_{s+1} - 2\,\mathbf{q}_s + \mathbf{q}_{s-1})$. Instead of (5.30), we use the full implicit position update formula $0 = \mathbf{q}_{s+1} - \mathbf{q}_s - \hat{\mathbf{u}}_s$, where $\mathbf{q}_{s+1} = \mathbf{q}(t_{s+1})$, $\mathbf{q}_s = \mathbf{q}(t_s)$ and $\mathbf{q}_{s-1} = \mathbf{q}(t_{s-1})$. According to this full implicit position update formula, the variables $\hat{\mathbf{u}}_E$ and $\hat{\mathbf{u}}_B$ in (5.29) are replaced by $\hat{\mathbf{u}}_{s+2} = \mathbf{q}_{s+2} - \mathbf{q}_{s+1}$ and $\hat{\mathbf{u}}_{s+1} = \mathbf{q}_{s+1} - \mathbf{q}_s$, which yields (5.29) to become $\mathbf{M}(\mathbf{q}_{s+1} - 2\,\mathbf{q}_s + \mathbf{q}_{s-1}) - \mathbf{h}_M\Delta t^2 - \mathbf{w}_M\hat{\mathscr{P}}$. This formula corresponds to the finite difference discretization multiplied by Δt^2.

5.1.4 Linearization of the Gap Function and Drift Stabilization

The nonlinearity of the gap function (5.3) causes the index-3 discretization to be nonlinear. An approach to handle this nonlinearity is to linearize the gap function. Instead of $g(\mathbf{q}_E) = g_E = 0$ the condition $\bar{g}_E = 0$ is used, in which $\bar{g}_E$ is a linear approximation of g_E,

$$\bar{g}_E = g_B + \frac{\partial g}{\partial \mathbf{q}}^{\top}\bigg|_B (\mathbf{q}_E - \mathbf{q}_B) = g_B + \mathbf{w}_B^{\top}(\mathbf{q}_E - \mathbf{q}_B). \tag{5.32}$$

Because $\bar{g}_E \neq g_E$ in general, such schemes might be affected from drift. Note that in (5.32) the term g_B appears explicit. The constraint is enforced by the implicit consideration of $\mathbf{q}_E$, i.e. by the term $\mathbf{w}_B^{\top}(\mathbf{q}_E - \mathbf{q}_B)$. This term is linear with respect to $\mathbf{q}_E$. The term $\mathbf{q}_E - \mathbf{q}_B$ can be replaced according to the position update formula by a term $\mathbf{u}_X\Delta t$ with $\mathbf{u}_X = \mathbf{u}_X(\mathbf{u}_E, \mathbf{u}_B)$. Division of (5.32) by Δt yields then a constraint on velocity level with an additional explicit drift stabilization $\frac{g_B}{\Delta t}$, i.e. the linearized index-3 problem yields a constraint stabilized index-2-like discretization. The approach is related to Baumgarte's stabilization method. Applying the linearization of the gap function on the discussed pendulum (5.1-5.3) yields

$$\mathbf{M}(\mathbf{u}_E - \mathbf{u}_B) = \mathbf{h}_M\Delta t - \mathbf{w}_B\hat{P}, \tag{5.33}$$

$$\mathbf{q}_E - \mathbf{q}_B = \Delta t\,\mathbf{u}_E, \tag{5.34}$$

$$0 = \frac{\sqrt{\mathbf{q}_B^{\top}\,\mathbf{q}_B} - L}{\Delta t} + \frac{\mathbf{q}_B^{\top}}{||\mathbf{q}_B||}\mathbf{u}_E = \frac{g_B}{\Delta t} + \gamma_E. \tag{5.35}$$

Figure 5.7 depicts the results of a simulation which uses (5.33-5.35) without the drift stabilization term $\frac{g_B}{\Delta t}$. Note that the scheme is fairly dissipative due to the pure implicit position update formula (5.34). Figure 5.8 shows the results of a simulation which uses the drift stabilization term $\frac{g_B}{\Delta t}$. The drift is reduced, but the fairly dissipative behaviour remains. Note the pure implicit form of the position update formula (5.34). As the drift stabilization term $\frac{g_B}{\Delta t}$ corresponds to a spring with stiffness $\frac{1}{\Delta t}$, oscillations may occur if the position update formula is chosen carelessly. The following relations should hold: If the constraint equation $\frac{g_B}{\Delta t} + \gamma_E = 0$ is fulfilled for the time steps s and $s+1$, i.e.

$$\frac{g_B^s}{\Delta t} + \gamma_E^s = 0, \qquad \frac{g_B^{s+1}}{\Delta t} + \gamma_E^{s+1} = 0, \tag{5.36}$$

then γ_E^{s+1} should be approximatively zero. Note that this is only the case if $g_B^{s+1} \approx g_B^s + \gamma_E^s \Delta t$, which corresponds to the fully implicit form of the position update formula. A discretization which has the above described properties but which uses the semi-explicit position update formula is

$$\mathbf{M}(\mathbf{u}_E - \mathbf{u}_B) = \mathbf{h}_M \Delta t - \mathbf{w}_M \hat{P}, \tag{5.37}$$

$$\mathbf{q}_E - \mathbf{q}_B = \Delta t \, \frac{\mathbf{u}_E + \mathbf{u}_B}{2}, \tag{5.38}$$

$$0 = \frac{g_B + \frac{1}{2}\mathbf{w}_M^\top \mathbf{u}_B \Delta t}{\Delta t} + \mathbf{w}_M^\top \mathbf{u}_E \approx \frac{g_M}{\Delta t} + \gamma_E. \tag{5.39}$$

Note that $g_B + \frac{1}{2}\mathbf{w}_M^\top \mathbf{u}_B \Delta t \approx g_M$, thus the drift stabilization term is evaluated in this case at the midpoint $\mathbf{q}_M$. The approach corresponds to the modified Θ-method of Jean [47], which will be discussed in Sect. 5.3.3. The results of a simulation which uses the discretization (5.37-5.39) are shown in Fig. 5.9. No drift is present, and the energy remains constant.

5.1.5 Conclusions

In the previous section different discretizations for differential algebraic equations of index 3 have been discussed. An index-1 formulation of an index-3 DAE should be avoided as it suffers from heavy drift in both positions and velocities. Index-2 formulations of an index-3 DAE are quite handy in application. Instead of the forces λ discrete percussions $\hat{\mathbf{P}} = \lambda \Delta t$ have to be used to avoid ill-conditioning. Unfortunately, also index-2 problems suffer from drift in the positions. If $\varepsilon \to 1$, then the drift becomes minimal for circle-like trajectories. Attention has to be paid that small oscillations due to numerical errors in the constraints are not damped if ε is chosen equal to one. When ε is chosen close to zero, a drift in the positions is apparent for larger step sizes and nonlinear trajectories. Three approaches for index-3 discretizations have been discussed. The first approach is the preconditioning method of Botasso, which avoids ill-conditioning by using the variables $\hat{\mathbf{P}}\Delta t$ and $\mathbf{u}\Delta t$ to solve the problem. The GGL approach uses an index-2 formulation of the index-3 DAE together with a projection to guarantee admissible positions. Note that this

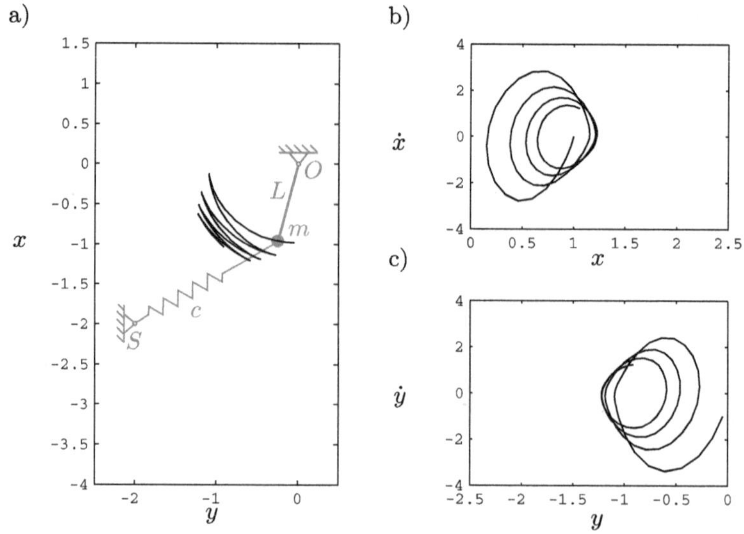

Fig. 5.7 Simulation without drift stabilization term $\frac{g_B}{\Delta t}$. Figure **a** visualizes the path of the system in the x-y-plane, the Figs. **b** and **c** depict the phase plots x versus $\dot{x}$ and y versus $\dot{y}$. Drift and an energy loss are obvious

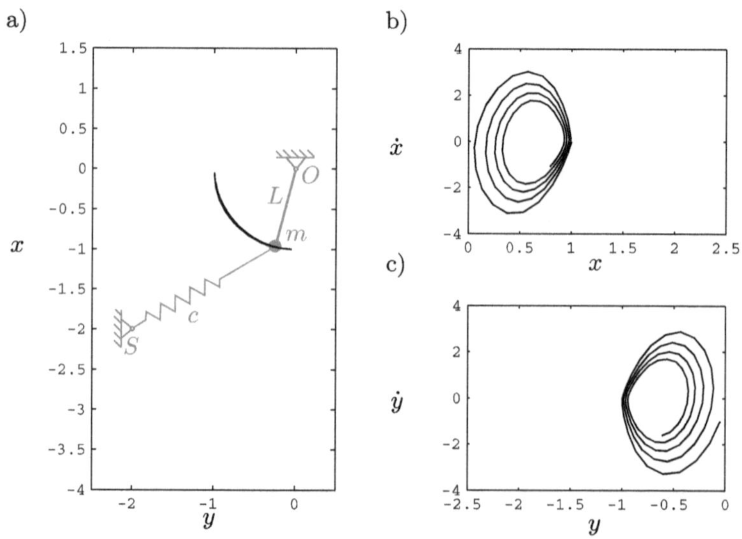

Fig. 5.8 Simulation with drift stabilization term $\frac{g_B}{\Delta t}$. Figure **a** visualizes the path of the system in the x-y-plane, the Figs. **b** and **c** depict the phase plots x versus $\dot{x}$ and y versus $\dot{y}$. The drift is prevented, the energy decrease remains

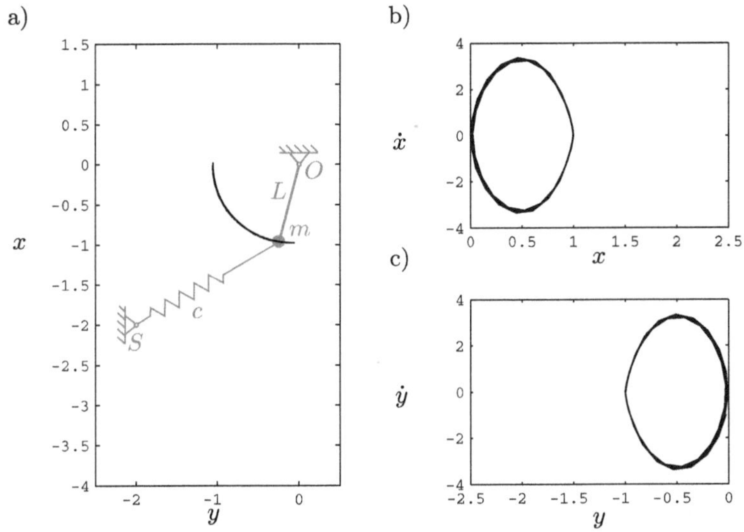

Fig. 5.9 Simulation with drift stabilization term $\frac{g_B+\frac{1}{2}\mathbf{w}_M^\top \mathbf{u}_B\Delta t}{\Delta t}$ and semi-explicit position update formula. Figure **a** visualizes the path of the system in the x-y-plane, the Figs. **b** and **c** depict the phase plots x versus $\dot{x}$ and y versus $\dot{y}$

projection might increase the energy of the system. Another approach linearizes the gap function, which yields an index-2-like discretization with an additional drift stabilization. Attention has to paid when defining the position update formula in this case, as careless choices may induce oscillations.

5.2 Time Evolution of Non-smooth Systems

The time evolution of non-smooth systems can be separated into piecewise smooth parts and switching points. As discussed in Sect. 3.2.1, a set-valued law can either be in constraint mode, i.e. it enforces a constraint equation, or in impressed mode. We gather this information in a vector $\sigma = \sigma(t)$ which indicates the state of the set-valued laws. If the entry of the i-th set-valued law σ_i is equal to 0, then the i-th set-valued law is assumed to be in constraint mode. If $\sigma_i = 1$, then the i-th set-valued law is in impressed mode. A smooth interval $[t_0, t_1]$ is characterized by a state $\sigma(t)$ which is constant for $t \in [t_0, t_1]$. Such a smooth interval can be described by a conventional DAE. Note that each possible state σ has an associated DAE, which we call also the underlying DAE. Switching points are time instances t_S at which the state σ changes, i.e. $\sigma^-(t_S) \neq \sigma^+(t_S)$.

Example

Consider a planar model of a point mass and a table. The position of the point mass is described by the coordinates $\mathbf{q} = (x\, y)^{\top}$, where the x-axis points along the table surface $y = 0$. The admissible positions of the point mass are given by the condition $y \geq 0$. The frictional unilateral contact between point mass and table is modeled by two set-valued laws, i.e. by a unilateral contact and and by a planar friction element. The non-smooth system is described by

$$\begin{aligned} 0 &= \begin{pmatrix} m & 0 \\ 0 & m \end{pmatrix} \begin{pmatrix} \ddot{x} \\ \ddot{y} \end{pmatrix} - \begin{pmatrix} 0 & 1 \\ 1 & 0 \end{pmatrix} \begin{pmatrix} \lambda_1 \\ \lambda_2 \end{pmatrix}, \\ -y &\in \mathscr{N}_{\mathbb{R}_0^+}(\lambda_1) \\ -\dot{x} &\in \mathscr{N}_{\mathscr{S}_p}(\lambda_2), \qquad \mathscr{S}_p = [-\mu\lambda_1, \mu\lambda_1]. \end{aligned} \tag{5.40}$$

Table 5.1 lists the associated DAE's of the non-smooth system. The switching points are the time instances at which an impact, a take-off or a stick-slip transition occurs. At these switching points the underlying differential algebraic equation changes. Note that an impact between point mass and table must be evaluated by a separate impact law in order to obtain the post-impact velocities, which serve as new initial conditions for the further movement. Thus an impact is not only a switching point at which the underlying DAE changes but also a time instance at which the states $\mathbf{q}$ and $\mathbf{u}$ of the system are re-initialized.

Special attention has to be paid for accumulative switching points. At such points, infinitely many switching points occur in a finite time. Take for instance a point mass which falls on a table. After each impact on the table the point mass will jump less high and the flight time becomes less long. There will be an accumulation of

Table 5.1 DAE's associated with (5.40)

state σ	equations of motion and algebraic constraints
$(1\ 1)^{\top}$ free flight	$0 = \begin{pmatrix} m & 0 \\ 0 & m \end{pmatrix} \begin{pmatrix} \ddot{x} \\ \ddot{y} \end{pmatrix}$
$(0\ 1)^{\top}$ closed, sliding	$0 = \begin{pmatrix} m & 0 \\ 0 & m \end{pmatrix} \begin{pmatrix} \ddot{x} \\ \ddot{y} \end{pmatrix} - \begin{pmatrix} 0 & 1 \\ 1 & 0 \end{pmatrix} \begin{pmatrix} 1 \\ -\frac{\dot{x}}{\lvert\dot{x}\rvert}\mu \end{pmatrix} \lambda_1, \quad y = 0$
$(0\ 0)^{\top}$ closed, sticking	$0 = \begin{pmatrix} m & 0 \\ 0 & m \end{pmatrix} \begin{pmatrix} \ddot{x} \\ \ddot{y} \end{pmatrix} - \begin{pmatrix} 0 & 1 \\ 1 & 0 \end{pmatrix} \begin{pmatrix} \lambda_1 \\ \lambda_2 \end{pmatrix}, \quad y = 0,\ \dot{x} = 0$
$(0\ 1)^{\top}$ free flight, sliding,	does not occur

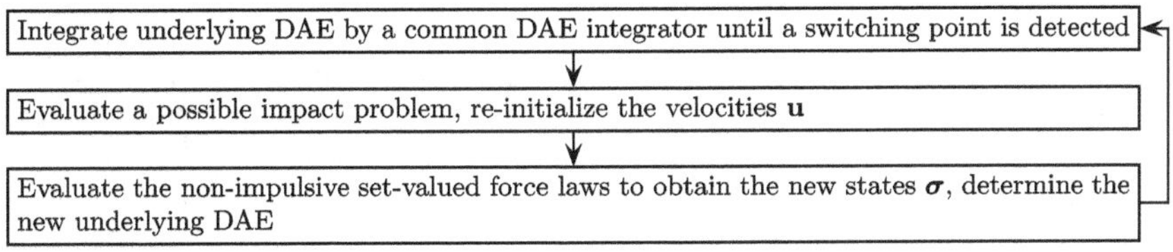

Fig. 5.10 Flow chart of an event-driven scheme

switching points, i.e. infinitely many impacts in a finite time, by the end of the point mass's movement.

5.2.1 Event-Driven Schemes

A first approach to simulate non-smooth systems are the event-driven methods [39, 57, 73]. These methods separate the motion of a non-smooth system into piecewise smooth parts and in switching points. The underlying DAE of a piecewise smooth part is integrated with an arbitrary DAE integrator until a switching point occurs. At this time instance, the set-valued impact and force laws (3.36) and (3.11-3.13) are evaluated and the new underlying DAE is identified. Note that a possible impact problem must be solved prior to the evaluation of the non-impulsive behaviour, as this evaluation must be with respect to the post-impact velocities, which serve also as new initial conditions for the further integration. Event-driven methods clearly distinguish between classical smooth integration and evaluation of the set-valued laws at the switching points. The methods have successfully been implemented for systems with few set-valued laws. Event-driven methods are very accurate but bring a lot of administrative effort and are not suited for accumulative switching points, because the methods aim at resolving all switching points. Furthermore, event-driven methods request for each state σ an associated underlying DAE, which becomes a challenging problem if the number of set-valued laws is large. A flow chart of the event-driven algorithm is depicted in Fig. 5.10. To conclude, event-driven methods split the time evolution of the non-smooth system into a sequence of differential algebraic equations, and discretize these DAE in a classical sense. Event-driven methods require switching point detection, evaluation of the set-valued impact and the set-valued force laws at these points and a possible re-initialization of the system in the case of an impact. This work focusses not on even-driven but on time-stepping methods, which are discussed in the next Sect. 5.3.

5.3 Time-Stepping Methods

In contrast to event-driven schemes, time-stepping methods discretize directly the measure differential inclusions (3.39-3.40). Therefore, no switching point detection is required. For each time step, an appropriate discrete state $\hat{\sigma}$ is determined. This discrete state $\hat{\sigma}$ is assumed to hold over the entire time step. Consider for instance a unilateral contact: in each time step the algorithm decides whether the whole time step is computed by the assumption of a closed or of an open contact. As a consequence, the discrete scheme can only switch its discrete state $\hat{\sigma}$ at discrete time points, i.e. at the begin or end point of a time step. A small step size has to be chosen in order to resolve switching points properly. Note that only for small step sizes the evolution of σ and $\hat{\sigma}$ can become similar. Time-stepping methods do not distinguish between the different smooth parts and switching points, and are therefore very robust and simple. The methods are well suited for accumulative switching points, and simulations of systems with up to several thousand set-valued laws are possible. As a drawback, time-stepping methods require a very small step size, and the accuracy is less satisfactory.

Similar to the DAE's one may distinguish between index-2 and index-3 time-stepping schemes. Note that an index-1 time-stepping discretization does not make sense because impacts can not be handled on acceleration level. Therefore, this work focuses on index-2 or index-3 time-stepping schemes.

Special attention has to be paid to geometric set-valued laws within an index-2 time-stepping scheme. As discussed in Sect. 3.2.1, geometric set-valued laws can only be treated on velocity level by an inclusion if $\mathbf{g}_i = 0$. Otherwise they are in impressed mode, see also (3.46). In the discrete case, it is not easy to define whether a discrete representation of a geometric set-valued law can be treated on velocity level by an inclusion or not, as the condition $\mathbf{g}_i = 0$ is never exactly fulfilled. We refer to a discrete set-valued law, which can be expressed on velocity level by an inclusion, as an *active* set-valued law. Each index-2 time-stepping scheme must provide a suitable tool to decide whether a discrete geometric set-valued law is active or not. Kinematic set-valued laws do not cause problems as they can always be expressed on velocity level by an inclusion, i.e. discrete kinematic set-valued laws are always active.

The following sections give an overview on existing time-stepping schemes. *Note that the original notation of these time-stepping schemes is adapted to the notation used in this work. In addition, the methods have sometimes been written in a different way.* For the original methods, their application and their interpretation, the reader is referred to the original contributions. Furthermore, two new index-3 schemes are introduced, which extend the GGL and the preconditioning approach to time-stepping schemes. A discussion concludes the section.

5.3.1 General Discretization Technique

In the following, a general discretization technique is discussed, of which most of the time-stepping methods can be derived from. Consider a time step with step size Δt, of which begin and end point are denoted by the indices B and E, respectively. The equality of measures (3.39) is discretized to obtain

$$\begin{aligned}
&\int_{[t_B,t_E]} \mathbf{M}\,\mathrm{d}\mathbf{u} \;\rightarrow \mathbf{M}_D(\mathbf{u}_E-\mathbf{u}_B),\\
&\int_{t_B}^{t_E} \mathbf{h}\,\mathrm{d}t \quad\;\rightarrow \mathbf{h}_D\Delta t,\\
&\int_{[t_B,t_E]} \mathbf{W}\mathrm{d}\mathbf{P}_i \rightarrow \mathbf{W}_D\hat{\mathbf{P}}_i,\\
&\Rightarrow \quad \mathbf{M}_D\,(\mathbf{u}_E-\mathbf{u}_B)-\mathbf{h}_D\,\Delta t-\sum_{i=1}^{n}\mathbf{W}_{Di}\hat{\mathbf{P}}_i=0,
\end{aligned} \tag{5.41}$$

where $\mathbf{M}_D$, $\mathbf{h}_D$ and $\mathbf{W}_{Di}$ are some approximations of the mass matrix $\mathbf{M}$, of the vector of external and gyroscopic forces $\mathbf{h}$ and of the generalized force directions $\mathbf{W}_i$,which will be specified later by the specific time-stepping schemes. The term $\hat{\mathbf{P}}_i$ refers to the discrete percussions. A general position update formula is

$$\mathbf{q}_E=\mathbf{q}_B+\big((1-\alpha)\mathbf{u}_B+\alpha\mathbf{u}_E\big)\Delta t, \quad \alpha\in[0,1]. \tag{5.42}$$

Some time-stepping schemes write the approximation $\mathbf{h}_D$ of $\mathbf{h}$ as

$$\mathbf{h}_D=(1-\beta)\mathbf{h}_B+\beta\bar{\mathbf{h}}_E, \quad \beta\in[0,1], \tag{5.43}$$

where $\mathbf{h}_B=\mathbf{h}(\mathbf{q}_B,\mathbf{u}_B,t_B)$. Furthermore, $\bar{\mathbf{h}}_E$ follows by linearization of $\mathbf{h}=\mathbf{h}(\mathbf{q},\mathbf{u},t)$,

$$\bar{\mathbf{h}}_E=\mathbf{h}_{B^\star}+\left.\frac{\partial\mathbf{h}}{\partial\mathbf{q}}\right|_{B^\star}(\mathbf{q}_E-\mathbf{q}_B)+\left.\frac{\partial\mathbf{h}}{\partial\mathbf{u}}\right|_{B^\star}(\mathbf{u}_E-\mathbf{u}_B). \tag{5.44}$$

Note that $\mathbf{h}$ and its partial derivatives with respect to $\mathbf{q}$ and $\mathbf{u}$ are evaluated for $\mathbf{q}_B$, $\mathbf{u}_B$ and t_E, i.e. $\mathbf{h}_{B^\star}=\mathbf{h}(\mathbf{q}_B,\mathbf{u}_B,t_E)$ etc. Note that the positions $\mathbf{q}$ and the velocities $\mathbf{u}$ are evaluated at the beginning, the time t at the end of the time step. Elimination of $\mathbf{q}_E-\mathbf{q}_B$ in (5.44) by the position update formula (5.42) gives

$$\bar{\mathbf{h}}_E=\mathbf{h}_{B^\star}+\left.\frac{\partial\mathbf{h}}{\partial\mathbf{q}}\right|_{B^\star}\big((1-\alpha)\mathbf{u}_B+\alpha\mathbf{u}_E\big)\Delta t+\left.\frac{\partial\mathbf{h}}{\partial\mathbf{u}}\right|_{B^\star}(\mathbf{u}_E-\mathbf{u}_B). \tag{5.45}$$

Inserting (5.45) in the discretized equality of measures (5.41) yields finally

$$0=\widehat{\mathbf{M}}(\mathbf{u}_E-\mathbf{u}_B)-\widehat{\mathbf{h}}\Delta t-\sum_{i=1}^{n}\mathbf{W}_i\hat{\mathbf{P}}_i, \tag{5.46}$$

in which $\widehat{\mathbf{M}}$ and $\widehat{\mathbf{h}}$ are the modified mass matrix $\mathbf{M}$ and the modified $\mathbf{h}$-vector,

Table 5.2 Management of set valued laws

set-valued law	mode	considered by
geometric (unilateral $g_i(\mathbf{q}_M) \leq 0$)	active	inclusion (5.52)
geometric (unilateral $g_i(\mathbf{q}_M) > 0$)	non-active	impressed force in **h**-vector
kinematic	always active	inclusion (5.52)

$$\widehat{\mathbf{M}} = \mathbf{M}_D + \beta\alpha\Delta t^2 \frac{\partial \mathbf{h}}{\partial \mathbf{q}}\bigg|_{B^\star} - \beta\Delta t \frac{\partial \mathbf{h}}{\partial \mathbf{u}}\bigg|_{B^\star}, \tag{5.47}$$

$$\widehat{\mathbf{h}} = \mathbf{h}_B + \beta(\mathbf{h}_{B^\star} - \mathbf{h}_B) + \beta\Delta t \mathbf{u}_B \frac{\partial \mathbf{h}}{\partial \mathbf{q}}\bigg|_{B^\star}. \tag{5.48}$$

The approximation of $\mathbf{h}_D$ according (5.43) brings advantages when integrating stiff systems, as at least the linear part of $\mathbf{h}$ is considered as implicit. Note that a pure explicit consideration of $\mathbf{h}$ would result in a loss of numerical stability for stiff systems.

5.3.2 *Moreau's Midpoint Rule*

Moreau's midpoint method [65] calculates a midpoint $\mathbf{q}_M = \mathbf{q}_B + \mathbf{u}_B \frac{\Delta t}{2}$, $t_M = t_B + \frac{\Delta t}{2}$ and discretizes the equality of measures as follows:

$$\mathbf{M}_M (\mathbf{u}_E - \mathbf{u}_B) - \mathbf{h}_M \Delta t - \sum_{i=1}^{n} \mathbf{W}_{Mi} \hat{\mathbf{P}}_i = 0, \tag{5.49}$$

$$\mathbf{M}_M = \mathbf{M}(\mathbf{q}_M, t_M), \quad \mathbf{h}_M = \mathbf{h}(\mathbf{q}_M, \mathbf{u}_B, t_M), \quad \mathbf{W}_{Mi} = \mathbf{W}_i(\mathbf{q}_M, t_M).$$

The position update formula is

$$\mathbf{q}_E = \mathbf{q}_B + \frac{\mathbf{u}_B + \mathbf{u}_E}{2} \Delta t, \tag{5.50}$$

which corresponds to $\alpha = \frac{1}{2}$. The inclusion (3.40) is discretized by

$$\int_{[t_B, t_E]} \mathrm{d}\gamma_i \rightarrow \gamma_{Ei} - \gamma_{Bi}, \qquad \int_{[t_B, t_E]} \mathrm{d}\mathscr{A}_i \rightarrow \hat{\mathscr{A}}_i, \qquad \int_{[t_B, t_E]} \mathrm{d}\mathbf{P}_i \rightarrow \hat{\mathbf{P}}_i, \tag{5.51}$$

which yields

$$-(\gamma_{Ei} + \varepsilon_i \gamma_{Bi}) \in \mathscr{N}_{\hat{\mathscr{A}}_i}(\hat{\mathbf{P}}_i). \tag{5.52}$$

Note that Moreau's midpoint rule is an index-2 time-stepping scheme as it formulates the set-valued laws on velocity level. The scheme calculates the gap functions $\mathbf{g}_i$ of all discrete geometric set-valued laws at the midpoint $\mathbf{q}_M$ in order to evaluate whether these are active or not. An active discrete set-valued law is considered by the inclusion (5.52). A non-active discrete set-valued law is in impressed mode and

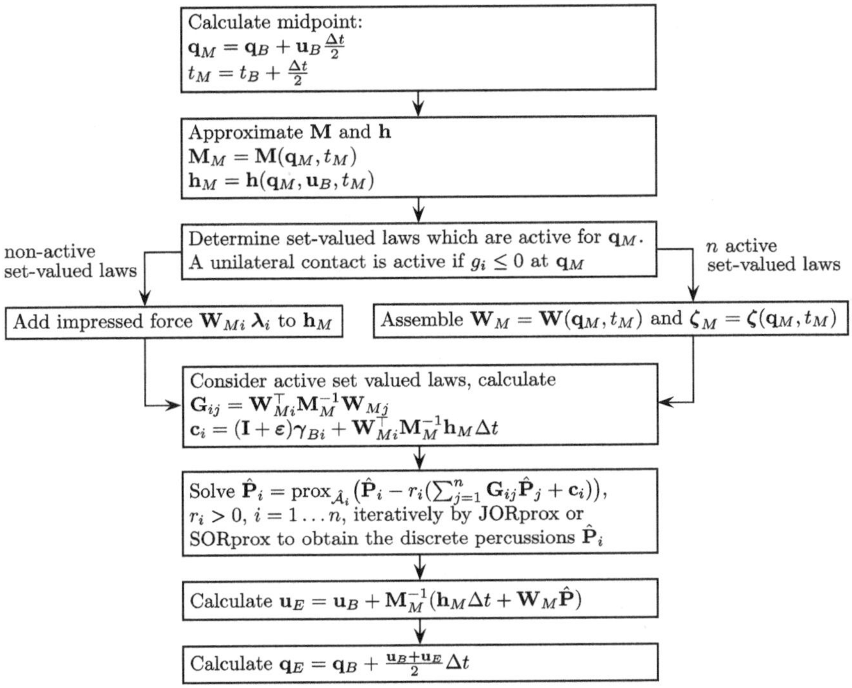

Fig. 5.11 Flow chart of a Moreau time step. Consider Sect. 3.1 for the definition of the generalized force directions $\mathbf{W}$ and the term ζ

must not be considered by an inclusion but by an external force which can be handled in the same way as the vector $\mathbf{h}$. Consider for example the unilateral contact. The contact is assumed active if $g_i \leq 0$. An open non-active unilateral contact $g_i > 0$ is neglected because its force λ_i is equal to zero. Geometric set-valued laws other than the unilateral contact require individual criteria to decide whether the law is active or not. Table 5.2 provides an overview on how discrete set-valued laws are managed.

Note the different states of the discrete set-valued law (5.52). Either the discrete percussion $\hat{\mathbf{P}}_i$ is chosen inside the set $\hat{\mathscr{A}}_i$ to enforce a restriction $\gamma_{Ei} + \varepsilon_i \gamma_{Bi} = 0$, or $\hat{\mathbf{P}}_i$ is impressed. Like in Sect. 3.2.1, a distinction between constraint and impressed mode can be made. Unlike Sect. 3.2.1, this distinction applies not only to one time instance but to a whole time step, i.e. the discrete set-valued law is assumed to act either as a constraint or as an impressed force during the entire time step, see also Table 5.3. Note that the constraint-like state $\gamma_{Ei} + \varepsilon \gamma_{Bi} = 0$ with $\varepsilon > 0$ must hold for both the impact and non-impact case. Considering the unilateral contact, an impact is assumed if γ_{Bi} is large enough such that $\gamma_{Ei} = -\varepsilon \gamma_{Bi}$ causes the contact to open again. On the other hand, if γ_{Bi} is as small that the contact remains in constraint mode, then the contact is sustaining, see also Table 5.4. Consider for example a point mass falling on a table with restitution coefficient $0 < \varepsilon < 1$. When the point

Table 5.3 Constraint and impressed mode of the discrete set-valued law (5.52). General case and example for unilateral contact

restriction	set-valued	state $\hat{\sigma}$ (entire time step)	Example unilateral contact ($g_M \leq 0$)	
$\gamma_{Ei} + \varepsilon_i \gamma_{Bi} = 0$	$\hat{\mathbf{P}}_i$	constraint ($\hat{\sigma}_i = 0$)	$\gamma_{Ei} + \varepsilon_i \gamma_{Bi} = 0$	$\hat{\mathbf{P}}_i \geq 0$
$\hat{\mathbf{P}}_i \in$ boundary of $\hat{\mathscr{A}}_i$	γ_{Ei}	impressed ($\hat{\sigma}_i = 1$)	$\hat{\mathbf{P}}_i = 0$	$\gamma_{Ei} \geq 0$

Table 5.4 Distinction whether the constraint $\gamma_{Ei} + \varepsilon_i \gamma_{Bi} = 0$, $\varepsilon > 0$ is associated with an impact or with a sustaining behaviour of the set-valued law. The index s denotes the number of the time step

$\gamma_{Ei} + \varepsilon_i \gamma_{Bi} = 0$	γ^s_{Bi}	$\hat{\sigma}^s_i$	$\hat{\sigma}^{s+1}_i$
impact	$\gamma^s_{Bi} >> 0$	0	1
sustaining behaviour	$\gamma^s_{Bi} \approx 0$	0	0

mass is not in contact with the table, then the force λ_i is zero and the set-valued law associated with the unilateral contact is in impressed mode. If the point mass touches the table, then the set-valued law changes to constraint mode and the velocity of the point mass is inverted according to $\gamma_{Ei} = -\varepsilon \gamma_{Bi}$. The contact will open again, the force λ_i will be zero and the set-valued law will be in impressed mode until the point mass touches the table again with a smaller velocity. After some elapsed time, the set-valued law will remain in constraint mode at a certain step. The relative velocity γ_i at the end of this time step will be so small that the set-valued law of the next time step will also be in constraint state. The discrete state $\hat{\sigma}_i$ does not change anymore, i.e. the mode of the set-valued law is sustaining. Note that the distinction between impact and sustaining behaviour is not made by the scheme (5.49-5.52), i.e. it treats impacts and sustaining behaviour with the same equation $\gamma_{Ei} + \varepsilon_i \gamma_{Bi} = 0$. To conclude, the discrete scheme (5.49-5.52) holds for both the smooth motion and the impact case, but only for small step sizes with satisfying accuracy.

In the following, the method (5.49-5.52) is rewritten in a convenient form which allows for solving the resulting inclusions by an augmented Lagrangian method [8, 15, 32, 57, 90, 93]. The relative velocities γ_i of the i-th active discrete set-valued law are linked to the generalized velocities $\mathbf{u}$ by $\gamma_i = \mathbf{W}_{Mi}^{\top}\mathbf{u} + \zeta_{Mi}$. By using this relation it is possible to transform the discrete scheme (5.49-5.52) into relative coordinates γ_i and $\hat{\mathbf{P}}_i$. Elimination of γ_i yields n inclusions which describe the individual behaviour of the n active discrete set-valued laws,

$$-(\sum_{j=1}^{n} \mathbf{G}_{ij}\hat{\mathbf{P}}_j + \mathbf{c}_i) \in \mathscr{N}_{\hat{\mathscr{A}}_i}(\hat{\mathbf{P}}_i). \tag{5.53}$$

The matrix $\mathbf{G}_{ij} = \mathbf{W}_{Mi}^{\top}\mathbf{M}_M^{-1}\mathbf{W}_{Mj}$ denotes the local Delassus matrix and $\mathbf{c}_i = (\mathbf{I}+\varepsilon)\gamma_{Bi} + \mathbf{W}_{Mi}^{\top}\mathbf{M}_M^{-1}\mathbf{h}_M\Delta t + \zeta_{Mi}$. The matrix ε is a diagonal matrix containing the restitution coefficients ε_i. Using the augmented Lagrangian approach, the inclusions (5.53) can be written as projective equations [90, 93],

$$\hat{\mathbf{P}}_i = \mathrm{prox}_{\hat{\mathscr{A}}_i}\Big(\hat{\mathbf{P}}_i - r_i\big(\sum_{j=1}^{n}\mathbf{G}_{ij}\hat{\mathbf{P}}_j + \mathbf{c}_i\big)\Big), \qquad r_i > 0. \tag{5.54}$$

Solving the n projective equations (5.54) with the JORprox or SORprox method yields the discrete percussion $\hat{\mathbf{P}}_i$, after which the generalized velocities $\mathbf{u}_E$ and the positions $\mathbf{q}_E$ can be calculated. A flow chart of a Moreau time step is depicted in Fig. 5.11.

5.3.3 The Modified Θ-Method

In [47] Jean presents the non-smooth contact dynamics method, which is also referred to as modified Θ method. Jean discretizes the equality of measures (3.39) to arrive at

$$0 = \mathbf{M}_B(\mathbf{u}_E - \mathbf{u}_B) - \big((1-\Theta)\mathbf{h}_B + \Theta\bar{\mathbf{h}}_E\big)\Delta t - \sum_{i=1}^{n}\mathbf{W}_{Bi}\hat{\mathbf{P}}_i, \tag{5.55}$$

$$\mathbf{q}_E = \mathbf{q}_B + \big((1-\Theta)\mathbf{u}_B + \Theta\mathbf{u}_E\big)\Delta t, \quad \Theta \in [0,1]. \tag{5.56}$$

Note that Θ corresponds to $\alpha = \beta$ in (5.42-5.43). The mass matrix $\mathbf{M}$ and the generalized force directions $\mathbf{W}_i$ are evaluated at an auxiliary intermediate value of $\mathbf{q}$, for example $\mathbf{q}_B$, i.e. $\mathbf{M}_B = \mathbf{M}(\mathbf{q}_B)$ etc. The method is unconditionally stable for $\Theta \geq \frac{1}{2}$ as far as linear systems are concerned. Friction is modeled according to $-\mathbf{W}_i^{\top}\mathbf{u}_E \in \mathscr{N}_{\mathscr{S}_i}(\hat{\mathbf{P}}_i)$, where the convex set $\mathscr{S}_i$ defines the admissible sliding forces. Jean discusses planar friction as well as spatial friction. A unilateral contact is modeled by

$$-\frac{g_{Di}}{\Delta t} + \gamma_{Ei} \in \mathscr{N}_{\mathbb{R}_0^+}(\hat{P}_i), \tag{5.57}$$

in which g_{Di} is an approximate gap associated with the position $\mathbf{q}_D = \mathbf{q}_B + (1-\Theta)\Delta t\,\mathbf{u}_B$. Note the connection between the proposed unilateral contact law (5.57) and (5.35,5.39). As only the velocities appear implicit, the time-stepping scheme corresponds to a drift stabilized index-2-like scheme. In the following, the motivation for using the proposed unilateral contact law (5.57) is discussed, see also Sect. 5.1.4. Let s and $s+1$ be two successive time steps. It holds that

$$\begin{aligned} g_{Di}^{s+1} &= g_{Bi}^{s+1} + (1-\Theta)\gamma_{Bi}^{s+1}\Delta t = g_{Ei}^{s} + (1-\Theta)\gamma_{Ei}^{s}\Delta t \approx \bar{g}_{Ei}^{s} + (1-\Theta)\gamma_{Ei}^{s}\Delta t \\ &= g_{Bi}^{s} + (1-\Theta)\gamma_{Bi}^{s}\Delta t + \gamma_{Ei}^{s}\Delta t = g_{Di}^{s} + \gamma_{Ei}^{s}\Delta t, \end{aligned} \tag{5.58}$$

with

$$\bar{g}_{Ei} = g_{Bi} + \mathbf{w}_B^\top(\mathbf{q}_E - \mathbf{q}_B) \approx g_{Bi} + \big((1-\Theta)\gamma_B + \Theta\gamma_E\big)\Delta t. \tag{5.59}$$

When using the proposed unilateral contact law, both the gap g_D and the relative velocity γ_E become approximatively zero within two time steps. A first time step with $g_{Di}^s + \gamma_{Ei}^s = 0$ yields $g_{Di}^{s+1} = 0$ by (5.58). Since $\gamma_E^s \leq 0$ due to $g_D^s \geq 0$, a further time step $g_{Di}^{s+1} + \gamma_{Ei}^{s+1} = 0$ is applied, which yields finally $\gamma_{Ei}^{s+1} = 0$. In order to consider inelastic impacts, Jean modifies the approximate gap function to become $g_{Di} \leftarrow \max(0, g_{Di})$. Note that for $g_{Di} = 0$, the unilateral contact law (5.57) becomes equal to an inelastic impact law on velocity level. It is quite important to realize that the unilateral contact law and the position update formula interact with each other. The Θ-method allows for arbitrary position update formulas and adjusts the unilateral contact laws accordingly. If the unilateral contact law is chosen independently from the position update formula, then the relative velocities γ_i do not vanish and the treatment of impacts is unpredictable. In addition, sustaining contacts may begin to oscillate due to numerical errors. A detailed discussion the Θ-method is given by Funk [36].

Jean solves the inclusion problem by a successive solution of all set-valued laws. While a one-dimensional set-valued law i associated with a unilateral contact or planar friction is solved straightforward as function of all other forces $\lambda_{i \neq j}$, the spatial friction problem is solved by iterating the associated projective λ_i-equation.

5.3.4 *Time-Stepping by Paoli and Schatzman*

Paoli and Schatzman propose in [68, 69] an index-3 time-stepping scheme which does not discretize the equality of measures but the equations of motion using finite differences. In [68] the one-dimensional case is discussed, i.e. $m\ddot{x} - h - \lambda = 0$ and $-x \in \mathcal{N}_{\mathbb{R}_0^+}(\lambda) \Leftrightarrow -\lambda \in \mathcal{N}_{\mathbb{R}_0^+}(x)$. The discretization yields

$$0 = \frac{x_{s+1} - 2x_s + x_{s-1}}{\Delta t^2} - h_s - \lambda, \tag{5.60}$$

$$-\lambda \in \mathcal{N}_{\mathbb{R}_0^+}\Big(\frac{x_{s+1} + \varepsilon x_{s-1}}{1+\varepsilon}\Big), \tag{5.61}$$

in which x_s is the value of x at t_s and $h_s = h(x_s, x_{s-1}, t_s, \frac{x_{s+1} - x_{s-1}}{2\Delta t})$. The parameter ε corresponds to the restitution coefficient. Paoli and Schatzman omit an ill-conditioning by elimination of λ,

$$\frac{x_{s+1} - 2x_s + x_{s-1}}{\Delta t^2} + \mathcal{N}_{\mathbb{R}_0^+}\Big(\frac{x_{s+1} + \varepsilon x_{s-1}}{1+\varepsilon}\Big) \ni h_s. \tag{5.62}$$

The authors transform the inclusion (5.62) into a projective equation, which is solved for x_{s+1}. Note that (5.61) incorporates an impact law. Consider two successive time steps in which the unilateral contact law (5.61) is in constraint mode, i.e. $x_{s+1} + \varepsilon x_{s-1} = 0$ and $x_{s+2} + \varepsilon x_s = 0$. Subtracting the first equation from the second yields $x_{s+2} - x_{s+1} = -\varepsilon(x_s - x_{s-1})$, which can be divided by Δt to obtain Newtons impact law. In [69] the authors extend the theory to the multi-dimensional case. Furthermore, the authors prove convergence of their scheme.

5.3.5 Time-Stepping by Stiegelmeyr, Funk, Foerg, Pfeiffer et al.

Stiegelmeyr [85] uses the discrete equality of measures (5.41) and the fully implicit position update formula (5.42) with $\alpha = 1$ and approximates $\mathbf{M}$, $\mathbf{h}$ and $\mathbf{W}_i$ at the beginning of the time step $(\mathbf{q}_B, \mathbf{u}_B, t_B)$. A unilateral contact is modeled by $-g_{Ei} = -(g_{Bi} + \gamma_{Ei}\Delta t) \in \mathscr{N}_{\mathbb{R}_0^+}(\hat{P}_i)$, which corresponds to a linearization of the gap function. The formulation incorporates an inelastic impact law. Furthermore, Stiegelmeyr suggests a second order discretization according to Heun, which is not applicable in case of impacts. Stiegelmeyr writes the inclusion problem as linear complementarity problem and gives some solution approaches. In particular, he proposes a semi-enumerative LCP-solver. Special attention has to be paid to partially elastic impacts. A time step with such an impact is rejected and recalculated using Moreau's discrete set-valued law (5.52), which is capable of both partially elastic impacts and non-impulsive motion.

Funk [36] uses the general formulation (5.42) and (5.43) together with the discrete set-valued law (5.57) proposed by Jean, which Funk discusses in detail. Different to Jean he writes the unilateral contact law scaled by Δt, i.e. $-(g_{Di} + \gamma_{Ei}\Delta t) \in \mathscr{N}_{\mathbb{R}_0^+}(\hat{P}_i)$. Funk performs the linearization (5.45) of $\mathbf{h}$ not only with respect to $\mathbf{q}$ and $\mathbf{u}$ but also with respect to the time t. Therefore, the scheme becomes slightly different to (5.47) and (5.48), i.e. $\mathbf{h}$ and its partial derivatives are evaluated at the begin point B and some additional terms $\frac{\partial \mathbf{h}}{\partial t}$ occur. Funk does a stability analysis considering a linear unconstrained system and provides stability plots for different choices of the parameters α and β. For linear stiff systems he suggests $\alpha = \beta = \frac{1}{2}$, hence this choice provides the most stable scheme, which corresponds to the results of Jean. Furthermore, Funk introduces a matrix $\mathbf{Q}$, which allows for using different parameterizations for $\mathbf{u}$ and $\mathbf{q}$, i.e. $\dot{\mathbf{q}} = \mathbf{Q}\mathbf{u}$ a.e. This notation is especially of interest when rigid body rotations are described by Euler parameters, see also Sect. 7.6.1 or [54]. Funk solves the inclusion problem by turning it into a linear complementarity problem, which he solves by Lemke's method. In order to treat planar friction, Funk proposes a modified Lemke's method, which evaluates in each solution step the different states of the discrete set-valued laws, i.e. sliding in positive and negative direction as well as sticking. In addition, Funk proposes a Block-Gauss-Seidel method to solve the inclusion problem. This approach corresponds to a successive solution of the set-valued laws. Partially elastic impacts are handled separately similar to Stiegelmeyr.

Foerg, Pfeiffer and Ulbrich propose in [33, 72] a time-stepping scheme with $\alpha = 1$. The authors evaluate $\mathbf{M}$, $\mathbf{h}$ and $\mathbf{W}_i$ either at the beginning $(\mathbf{q}_B, \mathbf{u}_B, t_B)$ or at the end $(\mathbf{q}_E, \mathbf{u}_E, t_E)$ of the time step, which yields either an explicit or implicit method. The inclusion problem is then solved by an iterative augmented Lagrangian approach; an implicit formulation of $\mathbf{M}$, $\mathbf{h}$ and $\mathbf{W}_i$ is possible within the iterative loop. As Stiegelmeyr, the authors model the unilateral contact by $-\bar{g}_{Ei} = -(g_{Bi} + \gamma_{Ei}\Delta t) \in \mathscr{N}_{\mathbb{R}_0^+}(\hat{P}_i)$. The scheme incorporates a fully inelastic impact law. The authors compare the different solution approaches, i.e. LCP formulation, augmented Lagrangian approach and a Block-Gauss-Seidel iterative technique, which corresponds to the SORprox scheme. In [33] the authors discuss stability for an unconstrained linear system. In [34] Foerg et al. also use Moreau's midpoint rule to simulate a valve train module. In [32] the authors investigate convergence of the augmented Lagrangian method and discuss strategies for choosing the factor r.

Note that the structure of the set-valued laws of the methods discussed above are all associated with the Θ-method of Jean. While Funk leaves the specification of Θ open, Stiegelmeyr and Foerg et al. choose $\Theta = 1$. Unlike Jean, the discrete set-valued laws are not scaled with $\frac{1}{\Delta t}$, which yields the schemes to be ill-conditioned if only working with the discrete percussions $\hat{\mathbf{P}}$. The authors do acknowledge this fact by formulating the associated equations using scaled discrete percussions $\Delta t\, \hat{\mathbf{P}}$.

5.3.6 *Time-Stepping by Anitescu, Potra, Stewart, Trinkle et al.*

The group Anitescu, Potra, Stewart, Trinkle et al. developed in the recent years various time-stepping schemes, whose convergence has been proven in detail. All these schemes do not use an augmented Lagrangian approach but an LCP-formulation to solve the resulting inclusion problems. In the following, a brief overview on the different time-stepping methods is given. Note that the frictional unilateral contact used by Anitescu et al. is split in the following into a unilateral contact and a friction element.

5.3.6.1 Spatial Friction in LCP form

Anitescu, Potra, Stewart, Trinkle et al. formulate the set-valued laws as LCP. A unilateral contact i on displacement or on velocity level is represented by the complementarity $0 \preceq \lambda_i \perp g_i \succeq 0$ and $0 \preceq \lambda_i \perp \gamma_i \succeq 0$, respectively. A spatial friction element is more complicated. Anitescu et al. replace the set $\mathscr{S}_s = \{\boldsymbol{\lambda} \in \mathbb{R}^2 \mid ||\boldsymbol{\lambda}|| < a\}$ by an polyhedral approximation $\hat{\mathscr{S}}_s = \{\boldsymbol{\lambda} = \mathbf{D}\boldsymbol{\beta} \mid \boldsymbol{\beta} \geq 0, \mathbf{e}^\top \boldsymbol{\beta} < a\}$, where $\mathbf{e} = (1, \ldots, 1)^\top$, see [40, 52, 84]. The matrix $\mathbf{D} = (\mathbf{d}_1, \ldots, \mathbf{d}_k)$ consists of unit vectors which define $\hat{\mathscr{S}}_s$, where each $\mathbf{d}_i$ has an associate $\mathbf{d}_j$ with $\mathbf{d}_i = -\mathbf{d}_j$, see also Fig. 5.12. The associated LCP which approximates spatial friction is

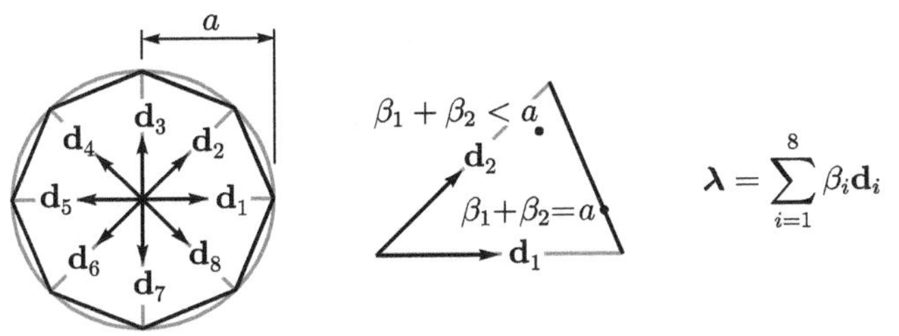

Fig. 5.12 Polyhedral approximation $\hat{\mathscr{S}}_s$ of $\mathscr{S}_s$

$$0 \preceq \beta \perp \lambda_S \mathbf{e} + \mathbf{D}^\top \mathbf{u} \succeq 0, \tag{5.63}$$
$$0 \preceq \lambda_S \perp a - \mathbf{e}^\top \beta \succeq 0, \tag{5.64}$$

where λ_S is a slack variable. Note that the relative velocity is zero if $\mathbf{D}^\top \mathbf{u} = 0$, i.e. $\mathbf{d}_i^\top \mathbf{u} = 0\, \forall i$. On the other hand, if at least one component $\mathbf{d}_i^\top \mathbf{u}$ of $\mathbf{D}^\top \mathbf{u}$ is positive, then there exists also a negative component $\mathbf{d}_j^\top \mathbf{u} = -\mathbf{d}_i^\top \mathbf{u}$, which enforces $\lambda_S > 0$ by (5.63). A detailed discussion on the formulation of spatial friction in LCP form is provided in [40, 84]. In the following, the discretization of the spatial friction element is given not in LCP form but as inclusions.

5.3.6.2 Stewart and Trinkle [84]

In [84] Stewart and Trinkle propose an index-3 time-stepping scheme

$$0 = \mathbf{M}_D(\mathbf{u}_E - \mathbf{u}_B) - \mathbf{h}_D \Delta t - \sum_{i=1}^{n} \mathbf{W}_i \hat{\mathbf{P}}_i, \tag{5.65}$$
$$\mathbf{q}_E = \mathbf{q}_B + \mathbf{u}_E \Delta t, \tag{5.66}$$

where $\mathbf{M}_D = \mathbf{M}(\mathbf{q}_B + \mathbf{u}_B \Delta t)$ and $\mathbf{h}_D = \mathbf{h}(\mathbf{q}_B + \mathbf{u}_B \frac{\Delta t}{2}, \mathbf{u}_B)$. The generalized force directions $\mathbf{W}_i$ are assumed to be constant. As a consequence, the gap function is linear with respect to $\mathbf{q}$. A unilateral contact is modeled on displacement level, $-(g_{Ei} - \delta_0) = -(\mathbf{w}_i \mathbf{q}_E - \delta_0) \in \mathscr{N}_{\mathbb{R}_0^+}(\hat{P}_i)$, where δ_0 is a penetration tolerance. Spatial friction is defined by $-\mathbf{W}_i^\top \mathbf{u}_E \in \mathscr{N}_{\mathscr{S}_{s,i}}(\hat{\mathbf{P}}_i)$. The proposed time-stepping scheme is only valid for inelastic collisions.

5.3.6.3 Anitescu and Potra [10]

Inspired by Stewart and Trinkle, Anitescu and Potra present in [10] an index-2 time-stepping scheme

$$0 = \mathbf{M}_D(\mathbf{u}_E - \mathbf{u}_B) - \mathbf{h}_D \Delta t - \sum_{i=1}^{n} \mathbf{W}_{Di} \hat{\mathbf{P}}_i, \qquad (5.67)$$

$$\mathbf{q}_E = \mathbf{q}_B + \Delta t\, \mathbf{u}_B, \qquad (5.68)$$

where $\mathbf{M}_D = \mathbf{M}(\mathbf{q}_B + \Delta t\, \mathbf{u}_B)$ and $\mathbf{W}_{Di} = \mathbf{W}_i(\mathbf{q}_B + \Delta t\, \mathbf{u}_B)$. The vector $\mathbf{h}_D$ is evaluated at the beginning, i.e. $\mathbf{h}_D = \mathbf{h}(\mathbf{q}_B, \mathbf{u}_B)$. Both, unilateral contact and spatial friction are described on velocity level, i.e. $-\mathbf{w}_i^\top \mathbf{u}_E \in \mathscr{N}_{\mathbb{R}_0^+}(\hat{P}_i)$ for i being a closed unilateral contact and $-\mathbf{W}_i^\top \mathbf{u}_E \in \mathscr{N}_{\mathscr{S}_{s,i}}(\hat{\mathbf{P}}_i)$ for i being a spatial friction element. The scheme also considers partially elastic impacts: it is proposed to estimate the collision time t_S, the collision position $\mathbf{q}_S$ and the pre-collision velocity $\mathbf{u}_S^-$, and to solve the impact at t_S by a Poisson impact law to obtain the post-impact velocity $\mathbf{u}_S^+$, see also Sect. 3.3 and [39]. Note that this procedure requests the solution of two additional impact inclusion problems. The integration is continued with the initial values t_S, $\mathbf{q}_S$ and $\mathbf{u}_S^+$. In the case of accumulative switching points it is suggested to set the Poisson restitution coefficient ε_p equal to zero for relative velocities less than a certain value. However, this is not applicable for all kind of accumulation points, consider for example the rocking rod [39].

5.3.6.4 Anitescu and Potra [11]

In [11] the authors present a linearly implicit time-stepping scheme for stiff systems. They linearize the vector of external and gyroscopic forces $\mathbf{h} = \mathbf{h}(\mathbf{q}, \mathbf{u})$ according to (5.44) and use the time-stepping scheme (5.41-5.42) with $\alpha = \beta = 1$,

$$0 = \mathbf{M}_B(\mathbf{u}_E - \mathbf{u}_B) - \bar{\mathbf{h}}_E \Delta t - \sum_{i=1}^{n} \mathbf{W}_{Bi} \hat{\mathbf{P}}_i, \qquad (5.69)$$

$$\mathbf{q}_E = \mathbf{q}_B + \mathbf{u}_E \Delta t. \qquad (5.70)$$

The mass Matrix $\mathbf{M}$ and the generalized force directions $\mathbf{W}_i$ are evaluated at the beginning, i.e. $\mathbf{M}_B = \mathbf{M}(\mathbf{q}_B)$ and $\mathbf{W}_{Bi} = \mathbf{W}_i(\mathbf{q}_B)$. Unilateral contacts and friction are modeled by $-\mathbf{w}_i^\top \mathbf{u}_E \in \mathscr{N}_{\mathbb{R}_0^+}(\hat{P}_i)$ and $-\mathbf{W}_i^\top \mathbf{u}_E \in \mathscr{N}_{\mathscr{S}_{s,i}}(\hat{\mathbf{P}}_i)$, respectively.

5.3.6.5 Anitescu and Hart [9]

In [9] Anitescu and Hart discuss a drift stabilization for the unilateral contact defined on velocity level. They propose to consider only unilateral contacts with $g_{Bi} < \delta_0$, where δ_0 is sufficient small. Such closed unilateral contacts are modeled by

$$-\frac{\bar{g}_{Ei}}{\Delta t} = -(\mathbf{w}_{Bi}^\top \mathbf{u}_E + \frac{g_{Bi}}{\Delta t}) \in \mathscr{N}_{\mathbb{R}_0^+}(\hat{P}_i), \qquad (5.71)$$

in which $\bar{g}_{Ei}$ is the linear approximation, see also (5.32). The position update formula is assumed to be $\mathbf{q}_E = \mathbf{q}_B + \mathbf{u}_E \Delta t$. The proposed time-stepping scheme evaluates $\mathbf{M}$, $\mathbf{h}$ and the $\mathbf{W}_i$'s at the begin point $(\mathbf{q}_B, \mathbf{u}_B, t_B)$. Note that this scheme is associated with the Θ-method with $\Theta = 1$, i.e. the method is associated with the work of Jean, Stiegelmeyr, Funk, Foerg and Pfeiffer et al.

5.3.6.6 Potra, Anitescu, Gravrea and Trinkle [74]

In [74] the authors present a time-stepping scheme with an integration order of two, which they call the linearly implicit trapezoidal method. The idea is to use a time stepping scheme which is of order two considering the piecewise smooth parts of the motion and to provide a reliable event (collision, take off, slip-stick) detection. In difference to event-driven methods, the behaviour of the system is never evaluated at acceleration level, but the time steps of the scheme are arranged such that they account for the position of the switching points. Potra et al choose the discretization (5.41-5.42) with $\alpha = \beta = \frac{1}{2}$. The mass matrix $\mathbf{M}$ as well as the generalized force directions $\mathbf{W}_i$ are approximated at the midpoint $\mathbf{q}_M = \mathbf{q}_B + \mathbf{u}_B \frac{\Delta t}{2}$, i.e. $\mathbf{M}_M = \mathbf{M}(t_M, \mathbf{q}_M)$ and $\mathbf{W}_{Mi} = \mathbf{W}_i(t_M, \mathbf{q}_M)$. The linear implicit trapezoidal time-stepping scheme yields

$$0 = \widehat{\mathbf{M}}(\mathbf{u}_E - \mathbf{u}_B) - \widehat{\mathbf{h}}\Delta t - \sum_{i=1}^{n} \mathbf{W}_{Mi}\hat{\mathbf{P}}_i, \tag{5.72}$$

where

$$\widehat{\mathbf{M}} = \mathbf{M}_M - \left.\frac{\partial \mathbf{h}}{\partial \mathbf{q}}\right|_{B^\star} \frac{\Delta t^2}{2} + \left.\frac{\partial \mathbf{h}}{\partial \mathbf{u}}\right|_{B^\star} \frac{\Delta t}{2}, \qquad \widehat{\mathbf{h}} = \left.\frac{\partial \mathbf{h}}{\partial \mathbf{q}}\right|_{B^\star} \frac{\Delta t^2}{2} + \frac{\Delta t}{2}(\mathbf{h}_B + \mathbf{h}_{B^\star}), \tag{5.73}$$

which corresponds to (5.47-5.48), i.e. the index $B^\star$ indicates a evaluation for $\mathbf{q}_B$, $\mathbf{u}_B$ and t_E. Instead of the partial derivatives $\left.\frac{\partial \mathbf{h}}{\partial \mathbf{q}}\right|_{B^\star}$ and $\left.\frac{\partial \mathbf{h}}{\partial \mathbf{u}}\right|_{B^\star}$ only approximations $\left.\frac{\partial \mathbf{h}}{\partial \mathbf{q}}\right|_{B^\star} + \mathscr{O}(\Delta t)$ and $\left.\frac{\partial \mathbf{h}}{\partial \mathbf{u}}\right|_{B^\star} + \mathscr{O}(\Delta t^2)$ are used. These approximations are chosen in a way that $\widehat{\mathbf{M}}$ remains positive definite. The unilateral contacts and friction are

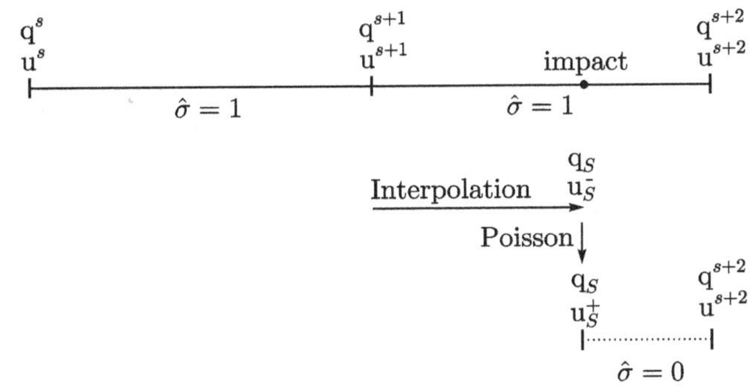

Fig. 5.13 Impact handling by Potra et al.

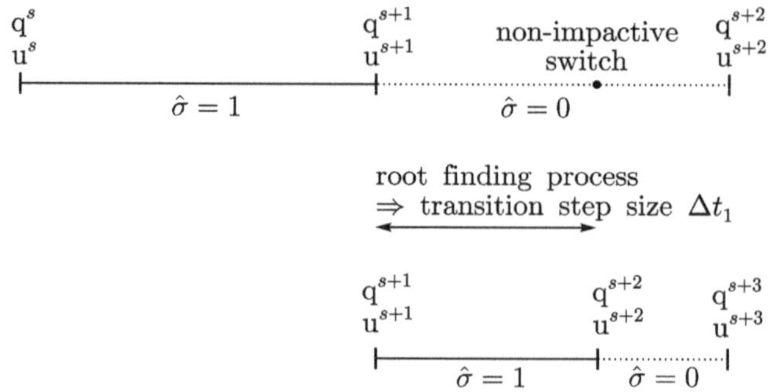

Fig. 5.14 Non-impactive switching point handling by Potra et al.

modeled by $-\mathbf{w}_{Mi}^{\top}\frac{\mathbf{u}_E+\mathbf{u}_B}{2} \in \mathscr{N}_{\mathbb{R}_0^+}(\hat{P}_i)$ and $-\mathbf{W}_{Mi}^{\top}\frac{\mathbf{u}_E+\mathbf{u}_B}{2} \in \mathscr{N}_{\mathscr{S}_{s,i}}(\hat{\mathbf{P}}_i)$, respectively. Only closed unilateral contacts according to $g_{Bi} \leq \max\{\delta_a, \delta_b \Delta t^3\}$ are considered, where δ_a and δ_b are user defined penetration tolerances. The authors show that the linearly implicit trapezoidal method has an integration order of two for the piecewise smooth motion. The time steps are arranged such that switching points are always located at a begin or end point. As a consequence, the discrete states $\hat{\sigma}$ can switch at the exact switching time instance, and the integration order is preserved. Switching points which occur when a geometric set-valued law switches from non-active to active, i.e. impacts associated with unilateral contacts, are handled by an event manager, which detects these switching points and uses cubic interpolation, i.e.

$$\mathbf{q}_{\mathrm{int}}(t) = \mathbf{q}_B,\ \mathbf{q}_{\mathrm{int}}(t+\Delta t) = \mathbf{q}_E,\ \mathbf{u}_{\mathrm{int}}(t) = \mathbf{u}_B,\ \mathbf{u}_{\mathrm{int}}(t+\Delta t) = \mathbf{u}_E, \tag{5.74}$$

to determine the impact time t_S, the impact position $\mathbf{q}_S$ and the impact velocity $\mathbf{u}_S^-$ of the first occurring impact. The post-impact velocity $\mathbf{u}_S^+$ is determined by the Poisson impact law, which requests solving two LCP's. The integration is continued at t_S with $\mathbf{q}_S$ and $\mathbf{u}_S^+$ being the new initial states. The impact handling of Potra et al. is summarized in Fig. 5.13. A time step which contains an impact remains in the old discrete state $\hat{\sigma} = \sigma^-(t_S)$, because closed unilateral contacts are detected at $\mathbf{q}_B$. Therefore, the obtained order two time evolution within this time step is valid up the time of switching. Potra et al. use cubic interpolation to determine the pre-impact states $\mathbf{q}_S$, $\mathbf{u}_S^-$ and t_S. The impact is evaluated to obtain the post-impact velocities $\mathbf{u}^+$, and the integration is continued at this time instance. Doing so, the time-stepping scheme can switch its discrete state $\hat{\sigma}$ at the "exact" switching time instance. Note that the discrete set-valued law $-\mathbf{w}_{Mi}^{\top}\frac{\mathbf{u}_E+\mathbf{u}_B}{2} \in \mathscr{N}_{\mathbb{R}_0^+}(\hat{P}_i)$ incorporates a fully elastic impact law according to Moreau. However, this impact behaviour of the discrete set-valued law is not taken into account as impacts are treated separately.

Switching points on velocity level, i.e. stick-slip transitions or take-off, are located by finding the root of some LCP specific functions, see Fig. 5.14. Using the augmented Lagrangian approach, this should correspond with finding the transition step size Δt_1 for which the $\mathrm{prox}_{\hat{\mathscr{A}}_i}$ switches its discrete state $\hat{\sigma}$, i.e. the transition step size Δt_1 for which

$$\hat{\mathbf{P}}_i - r_i(\sum_{j=1}^{n} \mathbf{G}_{ij}\hat{\mathbf{P}}_j + \mathbf{c}_i) \in \text{border of } \hat{\mathscr{A}}_i, \quad \mathbf{c}_i = \mathbf{c}_i(\Delta t). \tag{5.75}$$

A time step which contains the non-impactive switching point is rejected and repeated with a transition step size Δt_1, which yields the time step to end exactly at the switching time instance. Thus, the discrete state $\hat{\sigma}$ can switch at the "exact" switching time instance. Two problems remain unaddressed from our point of view. When calculating the transition time Δt, the "true" switching point of the mechanical system and the switching point of the discrete system must coincide, which is not always the case when using a discrete set-valued law $-\mathbf{W}_{Mi}^{\top}\frac{\mathbf{u}_E+\mathbf{u}_B}{2} \in \mathscr{N}_{\mathscr{S}_{s,i}}(\hat{\mathbf{P}}_i)$, see also Fig. 6.3. In addition, the repeated time step and the calculation of the transition time Δt_1 are based on different generalized force directions $\mathbf{W}_M$, as the step size and thus the midpoints $\mathbf{q}_M$ are different. This might eventually lead to contradictions.

5.3.6.7 Gavrea, Anitescu and Potra [37]

In [37] Gavrea et al. use the general discretization (5.41-5.43) with slight modifications[1]. The mass matrix $\mathbf{M}$ and the generalized force directions $\mathbf{W}$ are evaluated at the beginning of the time step, i.e. $\mathbf{M}_B = \mathbf{M}(\mathbf{q}_B, t_B)$ and $\mathbf{W}_{Bi} = \mathbf{W}(\mathbf{q}_B, t_B)$. The vector $\mathbf{h}$ is split into Coriolis terms $\mathbf{F}\mathbf{u}$ with $\mathbf{F} = \mathbf{F}(\mathbf{u})$, and into external forces $\mathbf{k}$, i.e. $\mathbf{h} = \mathbf{F}\mathbf{u} + \mathbf{k}$. The Coriolis terms are approximated as $\mathbf{F}_B((1+\beta)\mathbf{u}_B + \beta\mathbf{u}_E)$ with $\mathbf{F}_B = \mathbf{F}(\mathbf{u}_B)$. The external forces k are approximated by $(1-\beta)\mathbf{k}_B + \beta\mathbf{k}_E$ according to (5.44), i.e.

$$\bar{\mathbf{k}}_E = \mathbf{k}_{B^\star} + \frac{\partial \mathbf{k}}{\partial \mathbf{q}}\Big|_{B^\star}(\mathbf{q}_E - \mathbf{q}_B) + \frac{\partial \mathbf{k}}{\partial \mathbf{u}}\Big|_{B^\star}(\mathbf{u}_E - \mathbf{u}_B), \tag{5.76}$$

where $\mathbf{k}$ and its partial derivatives are evaluated for $\mathbf{q}_B$, $\mathbf{u}_B$ and t_E, i.e. $\mathbf{k}_{B^\star} = \mathbf{k}(\mathbf{q}_B, \mathbf{u}_B, t_E)$ etc. The difference $\mathbf{q}_E - \mathbf{q}_B$ in (5.76) can be eliminated using the position update which yields finally the time-stepping scheme

$$0 = \widehat{\mathbf{M}}(\mathbf{u}_E - \mathbf{u}_B) - \widehat{\mathbf{h}}\Delta t - \sum_{i=1}^{n} \mathbf{W}_{Bi}\hat{\mathbf{P}}_i, \tag{5.77}$$

$$\widehat{\mathbf{M}} = \mathbf{M}_B - \beta\Delta t\Big(\mathbf{F}_B + \frac{\partial \mathbf{h}}{\partial \mathbf{u}}\Big|_{B^\star}\Big) - \beta\alpha\Delta t^2 \frac{\partial \mathbf{h}}{\partial \mathbf{q}}\Big|_{B^\star}, \tag{5.78}$$

$$\widehat{\mathbf{h}} = \Big(\mathbf{k}_B + \beta(\mathbf{k}_{B^\star} - \mathbf{k}_B)\Big) + \beta\,\Delta t\,\mathbf{u}_B \frac{\partial \mathbf{k}}{\partial \mathbf{q}}\Big|_{B^\star} - (1-\alpha)\Delta t\,\mathbf{F}_B\mathbf{u}_B. \tag{5.79}$$

The unilateral contacts and spatial friction are modeled as $-\mathbf{w}_{Bi}^{\top}(\beta\mathbf{u}_E + (1-\beta)\mathbf{u}_B) \in \mathscr{N}_{\mathbb{R}_0^+}(\hat{P}_i)$ and $-\mathbf{W}_{Mi}^{\top}(\beta\mathbf{u}_E + (1-\beta)\mathbf{u}_B) \in \mathscr{N}(\mathscr{S}_{s,i})(\hat{\mathbf{P}}_i)$, respectively. Note that this

[1] They use for example the parameter γ instead of α and α instead of β. We keep the notation of (5.41-5.43).

formulation incorporates an impact law of Newton type. Nevertheless, the authors solve collisions by a Poisson impact law and event detection, as in [74].

5.3.7 Time-Stepping by GGL

In the following, the GGL formulation as discussed in Sect. 5.1.3.2 is extended to obtain a time-stepping scheme. Consider a point mass which falls on a table due to a gravitational force $F_g = mg$. The position and the velocity of the point mass is given by $x \geq 0$ and $u = \dot{x}$ a.e., respectively. The collision between point mass and table will be described by an impact law of Newton type. An associated GGL formulation yields

$$0 = m\,\mathrm{d}u + F_g \mathrm{d}t - \mathrm{d}P_\mathrm{u}, \tag{5.80}$$

$$\mathrm{d}P_\mathrm{u} \; : \; \begin{cases} x = 0 : -(u^+ + \varepsilon u^-) \in \mathcal{N}_{\mathbb{R}_0^+}(\mathrm{d}P_\mathrm{u}), \\ x > 0 : \mathrm{d}P_\mathrm{u} = 0, \end{cases} \tag{5.81}$$

$$0 = \mathrm{d}x - u\,\mathrm{d}t - \frac{1}{m}\,\mathrm{d}P_\mathrm{x}, \tag{5.82}$$

$$-(x + \delta) \in \mathcal{N}_{\mathbb{R}_0^+}(\mathrm{d}P_\mathrm{x}), \tag{5.83}$$

where $\delta \geq 0$ is a penetration tolerance. A possible discretization of (5.80-5.83) is

$$0 = m(u_E - u_B) + F_g \Delta t - \hat{P}_\mathrm{u}, \tag{5.84}$$

$$\hat{P}_\mathrm{u} \; : \; \begin{cases} x_M \leq 0 : -(u_E + \varepsilon u_B) \in \mathcal{N}_{\mathbb{R}_0^+}(\hat{P}_\mathrm{u}), \\ x_M > 0 : \hat{P}_\mathrm{u} = 0, \end{cases} \tag{5.85}$$

$$0 = x_E - x_B - \frac{u_E + u_B}{2}\Delta t - \frac{1}{m}\hat{P}_\mathrm{x}, \tag{5.86}$$

$$-(x_E + \delta) \in \mathcal{N}_{\mathbb{R}_0^+}(\hat{P}_\mathrm{x}), \tag{5.87}$$

where $x_M = x_B + \frac{\Delta t}{2} u_B$. The discrete equation of measures (5.84) and the discrete set-valued law on velocity level (5.85) allow for calculating the velocity u_E at the end of the time step. The approach corresponds to Moreau's midpoint rule, and both non-impulsive and partially elastic impact behaviour are included in the formulation. Elimination of u_E yields the inclusion

$$x_M \leq 0 \quad : \quad -\Big(\frac{\hat{P}_\mathrm{u} - F_g \Delta t}{m} + (1 + \varepsilon) u_B\Big) \in \mathcal{N}_{\mathbb{R}_0^+}(\hat{P}_\mathrm{u}), \tag{5.88}$$

of which the associated projective equation is

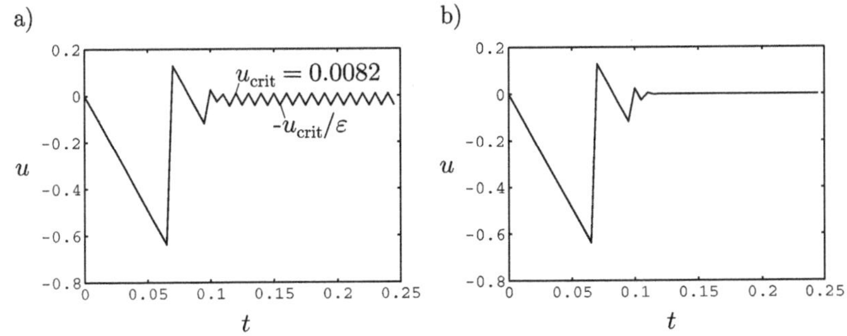

Fig. 5.15 Velocity of the point mass. In Fig. **a** oscillations due to $\delta = 0$ are apparent. Figure **b** shows the velocity for $\delta = 2.05\,10^{-5}$ m

$$x_M \leq 0 \quad : \quad \hat{P}_{\text{u}} = \text{prox}_{\mathbb{R}_0^+}(F_g\Delta t - (1+\varepsilon)mu_B). \tag{5.89}$$

The velocity u_E follows from (5.84). The position x_E is calculated from the position update formula (5.86) and the associated set-valued law on displacement level (5.87). In contrast to the position update formula (5.50) used by Moreau, the formula (5.86) contains a free parameter $\hat{P}_{\text{x}}$ which corresponds to a displacement Δx. This displacement is chosen such that the position becomes admissible, which corresponds to a projection. The associated projective equation is

$$\hat{P}_{\text{x}} = \text{prox}_{\mathbb{R}_0^+}(-m\,(x_B + \delta + \frac{u_E + u_B}{2}\Delta t)). \tag{5.90}$$

Note that such a projection can increase the energy of the system. Projections can also be problematic from another point of view. Consider the case $\delta = 0$, $x^s = 0$ and $u^s > 0$. The unilateral contact is open at the midpoint, i.e. $x^s + \frac{\Delta t}{2}u_s > 0$, which yields the velocity $u^{s+1} = u^s - \frac{F_g\Delta t}{m}$. This velocity is negative for small u^s, i.e. $u^{s+1} < 0$, which causes the contact to be closed in the next time step. In this case it holds that $u^{s+2} = -\varepsilon u^{s+1} = -\varepsilon(u^s - \frac{F_g\Delta t}{m}) > 0$. The position is $x^{s+2} = 0$ due to the projection. Note that for a certain critical velocity u_{crit} it holds that $u_{\text{crit}} = u^s = u^{s+2}$, which yields the states at t^{s+2} and t^s to be equal. The velocity will not decrease anymore because the energy decrease caused by the impact will be compensated by the energy increase due to the projection. The choice of the penetration tolerance δ is therefore essential. It must be chosen large enough, such that a critical velocity u_{crit} can not cause the unilateral contact to open at the midpoint, i.e. $-\delta + u_{\text{crit}}\frac{\Delta t}{2} \leq 0$. Such a choice is a sufficient but not a necessary condition. The critical velocity and the sufficient penetration tolerance for our example are

$$u_{\text{crit}} = \frac{F_g\Delta t}{m}\frac{\varepsilon}{1+\varepsilon}, \qquad \delta = u_{\text{crit}}\frac{\Delta t}{2}. \tag{5.91}$$

In Figs. 5.15 and 5.16 the velocity and the energy of the point mass are depicted. The following parameters have been used: $m = 1$ kg, $\Delta t = 0.005$s, and $\varepsilon = 0.2$. The

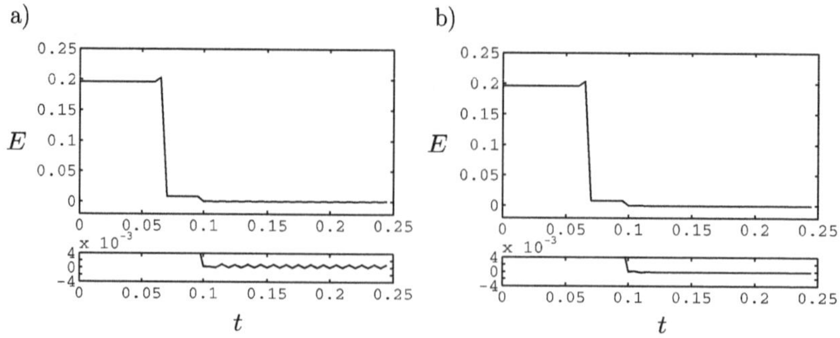

Fig. 5.16 Energy $E = \frac{1}{2}mu^2 + F_g x$ of the point mass. In Fig. **a** oscillations due to $\delta = 0$ are apparent. Figure **b** shows the energy for $\delta = 2.05\,10^{-5}$ m. Note the increase of energy at $t = 0.065$s due to the projection

Figs. 5.15a and 5.16a show the velocity and the energy for the case $\delta = 0$. The point mass oscillates between the velocities u_{crit} and $-u_{crit}/\varepsilon$ as each projection increases the energy which was lost by the impact. In Figs. 5.15b and 5.16b the penetration tolerance has been chosen $\delta = 2.05\,10^{-5}$ m. In this case the unilateral contact is mainly enforced by the set-valued law (5.85), and the projection associated with (5.87) is only applied if the penetration becomes large, see for example at $t = 0.065$s.

To conclude, the presented GGL time-stepping method consists of Moreau's mid-point rule with an additional projection, which guarantees admissible positions. Inelastic and elastic impacts can be handled. Due to the projection the energy of the system can increase. In addition, one has to allow for small penetrations as else oscillations might occur.

5.3.8 *Time-Stepping by Preconditioning*

In the following, the preconditioning approach as discussed in Sect. 5.1.3.3 is applied to non-smooth systems to obtain an index-3 time-stepping scheme. As in the previous section, the scheme is demonstrated for a point mass falling on a table, i.e.

$$0 = m\,\mathrm{d}u\,\mathrm{d}t + F_g \mathrm{d}t^2 - \mathrm{d}\mathscr{P}, \tag{5.92}$$

$$0 = \mathrm{d}x - u\,\mathrm{d}t, \tag{5.93}$$

$$-x \in \mathscr{N}_{\mathbb{R}_0^+}(\mathrm{d}\mathscr{P}). \tag{5.94}$$

Note that we do not formulate any impact conditions. A possible discretization of (5.92-5.94) yields

$$0 = m(\hat{u}_E - \hat{u}_B) + F_g \Delta t^2 - \hat{\mathscr{P}}, \tag{5.95}$$

$$0 = x_E - x_B - \big((1-\alpha)\hat{u}_B + \alpha \hat{u}_E\big), \tag{5.96}$$

$$-x_E \in \mathscr{N}_{\mathbb{R}_0^+}(\hat{\mathscr{P}}), \tag{5.97}$$

in which $\hat{u} = u\Delta t$. The position update formula (5.96) can be varied by the scalar $\alpha \in (0,1]$. Elimination of x_E and $\hat{u}_E$ in (5.95-5.97) yields finally

$$\hat{\mathscr{P}} = \text{prox}_{\mathbb{R}_0^+}\Big(-\frac{m}{\alpha}(x_B - \hat{u}_B) + F_g \Delta t^2\Big). \tag{5.98}$$

In case of an impact the discrete set-valued law associated with the unilateral contact is in constraint mode. The $\text{prox}_{\mathbb{R}_0^+}$ does not project on its set $\mathbb{R}_0^+$, i.e. $\hat{\mathscr{P}} = -\frac{m}{\alpha}(x_B - \hat{u}_B) + F_g \Delta t^2$, which yields together with (5.95)

$$\hat{u}_E = -\frac{1}{\alpha}x_B - \hat{u}_B(\frac{1}{\alpha} - 1) \quad \Rightarrow \quad u_E = -\frac{1}{\alpha \Delta t}x_B - u_B(\frac{1}{\alpha} - 1). \tag{5.99}$$

It appears that (5.99) plays the role of an impact law. The impact behaviour $u_B \rightarrow u_E$ is induced by the discretization, to be more precise by the position update formula. Unfortunately, equation (5.99) depends on the term $x_B/\Delta t$, which can not be neglected, not even for small x_B due to the division by Δt. Only the choice $\alpha = 1$ makes more or less sense. In this case it holds that $u_E = -\frac{x_B}{\Delta t}$ and $x_E = 0$ according to (5.99) and (5.96). The velocity u_E is still negative due to $x_B > 0$, thus the unilateral contact remains closed. Considering the next time step, $x_B^{(2)}$ is zero and therefore also the velocity $u_E^{(2)}$ becomes zero. Thus, the choice $\alpha = 1$ yields a fully inelastic impact within *two* time steps. Note once again the link to the modified Θ-method of Jean. In case of $\alpha < 1$ it is not guaranteed that u_E remains negative, and the unilateral contact may open again in the next time step. The choice of $\alpha < 1$ results in a partially elastic impact law (5.99) with hazardous post-impact velocities u_E, which depend not only on u_B but also on the non-negligible term $x_B/\Delta t$.

The discussed example is quite easy to solve because the gap function in (5.93) is linear. In order to show how the method is applied to more complex systems, a planar pendulum, which consists of a point mass m and an inelastic rope with length L, is simulated. The point mass is subjected to a gravitational force $F_g = mg$. Furthermore, the position and the velocity of the point mass are denoted by $\mathbf{q} = (x\,y)^\top$ and $\mathbf{u} = \dot{\mathbf{q}}$ a.e., respectively. The associated equality of measures is

$$0 = m\mathbf{I}\,\mathrm{d}\mathbf{u}\,\mathrm{d}t - (F_g\,0)^\top \mathrm{d}t^2 + \frac{\mathbf{q}}{||\mathbf{q}||}\mathrm{d}\mathscr{P} := \mathbf{M}\,\mathrm{d}\mathbf{u}\,\mathrm{d}t - \mathbf{h}\,\mathrm{d}t^2 - \mathbf{w}\mathrm{d}\mathscr{P}, \tag{5.100}$$

$$0 = \mathrm{d}\mathbf{x} - \mathbf{u}\,\mathrm{d}t, \tag{5.101}$$

$$-g = -(L - ||\mathbf{q}||) \in \mathscr{N}_{\mathbb{R}_0^+}(\mathrm{d}\mathscr{P}). \tag{5.102}$$

A possible discretization is

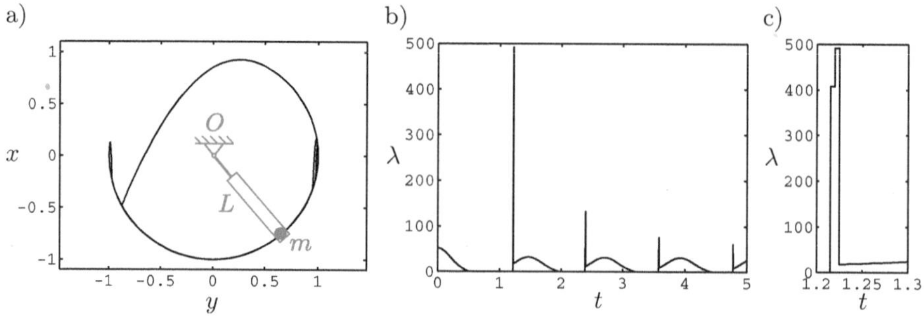

Fig. 5.17 Figure **a** shows the motion of the pendulum in the x-y-plane. Figures **b** and **c** plot the average forces $\lambda = \hat{\mathscr{P}}/\Delta t^2$

$$0 = \mathbf{M}(\hat{u}_E - \hat{u}_B) - \mathbf{h}\Delta t^2 - \mathbf{w}_M \hat{\mathscr{P}}, \tag{5.103}$$

$$0 = \mathbf{q}_E - \mathbf{q}_B - \hat{u}_E, \tag{5.104}$$

$$-g_E = -(L - ||\mathbf{q}_E||) \in \mathscr{N}_{\mathbb{R}_0^+}(\hat{\mathscr{P}}). \tag{5.105}$$

The Lagrange multiplier $\hat{\mathscr{P}}$ is determined by the iterative solution of

$$\hat{\mathscr{P}} = \mathrm{prox}_{\mathbb{R}_0^+}(\hat{\mathscr{P}} - r(L - ||\mathbf{q}_B + \hat{\mathbf{u}}_B + \mathbf{M}^{-1}(\mathbf{h}\Delta t^2 + \mathbf{w}_M \hat{\mathscr{P}}||)). \tag{5.106}$$

The results of the simulation are depicted in Fig. 5.17. Figure 5.17a shows the motion of the pendulum in the x-y-plane. Figures 5.17b and 5.17c show the average forces $\lambda = \hat{\mathscr{P}}/\Delta t^2$. Note that the impact is performed within two time steps, as can be seen in Fig. 5.17c. The following parameters have been used: $m = 1$ m, $L = 1$ m, $\Delta t = 0.005$s, $\mathbf{q}(t_0) = (-1\ 0)^\top$ m, $\mathbf{u}(t_0) = (0\ 6.5)^\top$ m/s.

5.3.9 *Discussion and Conclusions*

In the following, advantages, disadvantages and common properties of the presented time-stepping schemes are discussed.

5.3.9.1 Evaluation of M and $\mathbf{W}_i$

The various time-stepping schemes evaluate the mass matrix $\mathbf{M}$ and the generalized force directions $\mathbf{W}_i$ at an auxiliary intermediate position $\mathbf{q}$ and time t, for example $\mathbf{M}_M = \mathbf{M}(\mathbf{q}_M, t_M)$ etc. The specification of this state may influence the accuracy of method, but rarely the integration order. As the stability of the methods is determined on the linear unconstraint system with $\mathbf{M} = const$ [36], the choice of the approximate intermediate state has no influence on it. Moreau's midpoint evaluation is preferable, because it has shown good behaviour in application and because it comes closest to the second order trapezoidal approximation of $\mathbf{M}$ and $\mathbf{W}_i$.

5.3.9.2 Evaluation of h

The evaluation of $\mathbf{h}$ is crucial for stiff systems. Note that the stiffness originates mostly from $\mathbf{h}$, and that therefore an implicit consideration of $\mathbf{h}$ is advised in this case. Non-stiff systems can further on be calculated by an explicit choice of $\mathbf{h}$, for example $\mathbf{h}_M = \mathbf{h}(\mathbf{q}_M, \mathbf{u}_B, t_M)$. In order to avoid stiff terms in $\mathbf{h}$, stiff springs can also be modeled by a geometric set-valued force law (with a vanishing set-valued part, i.e. the law is always in impressed mode).

There are two ways to consider $\mathbf{h}$ in an implicit way. The first one is to use a linear approximation $\bar{\mathbf{h}}_E$ of $\mathbf{h}_E$, which yields a modified mass matrix $\widehat{\mathbf{M}}$ and a modified $\mathbf{h}$-vector $\widehat{\mathbf{h}}$, see (5.47-5.48). The choice $\alpha = \beta = \frac{1}{2}$ provides unconditional stability [36, 47] and yields in addition an order two scheme when choosing $\mathbf{M}_M = \mathbf{M}(\mathbf{q}_M, t_M)$ and $\mathbf{W}_{Mi} = \mathbf{W}_i(\mathbf{q}_M, t_M)$ [74]. Schemes which consider a linear approximation $\bar{\mathbf{h}}_E$ are called linearly implicit schemes. They have the same structure as explicit schemes and can be solved by the same methods. The only drawback when using a linear approximation $\bar{\mathbf{h}}_E$ of $\mathbf{h}_E$ is the need of the partial derivatives of $\mathbf{h}$ with respect to $\mathbf{q}$ and $\mathbf{u}$. These must either be provided by the user in an analytic way, or the algorithm has to calculate them numerically. In addition, one has to pay attention to not destroy the positive definiteness of the mass matrix when modifying it.

When using an iterative technique to solve the non-smooth problem, an implicit evaluation of $\mathbf{h}$ can be nested within the iterative loop. Take for instance the SORprox or JORprox method. The original iteration is done solely for the discrete percussions $\hat{\mathbf{P}}_i$. Note that for each update $\hat{\mathbf{P}}^\nu$, the corresponding velocities $\mathbf{u}_E^\nu$ and displacements $\mathbf{q}_E^\nu$ can be calculated, which yields $\mathbf{h}_E^\nu = \mathbf{h}(\mathbf{q}_E^\nu, \mathbf{u}_E^\nu, t_E)$. Thus the approximation $\mathbf{h}_D = \mathbf{h}_E^\nu$ yields a fully implicit scheme with respect to $\mathbf{h}$. Note that it is also possible to consider only the velocities in an implicit way, i.e. $\mathbf{h}_E^\nu(\mathbf{q}_M, \mathbf{u}_E^\nu, t_M)$, which avoids calculating position updates in each iterative step.

5.3.9.3 Formulations on Displacement and Velocity Level

The presented time-stepping schemes consider the discrete geometric set-valued law either on displacement or on velocity level, which yields an index-3 or an index-2 problem.

Formulations on Displacement Level

Discrete set-valued laws which are formulated on displacement level are not subjected to drift. In exchange, there are problems regarding the ill-conditioning of the associated DAE, the nonlinearity of the gap function and the treatment of impacts. A formulation on displacement level yields either an index-3 problem (preconditioning approach), a drift stabilized linear index-3 discretization (Stiegelmeyr, Funk, Foerg, Pfeiffer), a drift stabilized index-2-like discretization (Jean, Anitescu) or an index-2 problem with an additional projection to the admissible manifold $\mathbf{g} = 0$ (GGL approach).

A first problem is the nonlinearity of the gap function g of a unilateral contact. There are two ways to handle this nonlinearity. A first one is to linearize the gap function, as it is done by Jean [47], Stiegelmeyr [85], Funk [36], Foerg [33], Pfeiffer [72] and Anitescu [9], see also Sect. 5.1.4. Either one uses $-(\frac{\bar{g}_{Di}}{\Delta t} + \gamma_{Ei}) \in \mathscr{N}_{\mathbb{R}_0^+}(\hat{P}_i)$ to obtain a drift stabilized index-2-like problem (Jean, Anitescu et al), or one models a unilateral contact according to $-(\bar{g}_{Di} + \gamma_{Ei}\Delta t) \in \mathscr{N}_{\mathbb{R}_0^+}(\hat{P}_i)$ (Stiegelmeyr, Foerg). In the latter case, preconditioning has to be considered to avoid ill-conditioning. This procedure might be preferable, as the division of g_D by Δt might cause problems if $\Delta t \to 0$. The schemes formulate the discrete set-valued laws implicitly in the velocities and explicitly in the displacements. Therefore, a combination of linearized discrete set-valued laws on displacement level and discrete set-valued laws on velocity level is possible. Attention has to be paid to carefully choose the position update formula. As the linearization approach corresponds to a drift stabilized scheme, oscillations may occur if the criterions of Jean are not fulfilled, i.e. a discrete geometric set-valued force law in constraint mode must cause the relative velocity to become zero. While Jean and Funk use a general formulation, Stiegelmeyr, Foerg and Pfeiffer et al. restrict themselves to a fully implicit position update formula, which allows for choosing $g_D = g_B$. Also the drift stabilized scheme of Anitescu and Hard must be assigned to the latter category.

A second way to deal with the nonlinearity of $\mathbf{g}_i$ is to include these gap functions $\mathbf{g}_i$ into the projective equations, i.e.

$$-\mathbf{g}_E \in \mathscr{N}_{\hat{\mathscr{A}}_i}(\hat{\mathbf{P}}_i) \quad \Leftrightarrow \quad \hat{\mathbf{P}}_i = \mathrm{prox}_{\hat{\mathscr{A}}_i}(\hat{\mathbf{P}}_i - r_i \mathbf{g}_E). \tag{5.107}$$

This procedure is of course not possible when solving the non-smooth problem by a linear complementarity problem.

The treatment of impacts is closely connected to the position update formula. Note that conditions on displacement level can only influence the velocities via the position update formula, i.e. when stating discrete set-valued laws on displacement level one has to consider that the position update formula and thus the time discretization causes an impact like behaviour. There are two ways to deal with impacts. Either one models impacts within the discretization, or one detects and processes them by a separate impact evaluation.

Most formulations on displacement level are only capable to handle fully inelastic impacts. All schemes which use a linearization of the gap function, i.e. schemes with drift stabilization, request both gap function g and relative velocity γ to be zero for discrete geometric set-valued laws in constraint mode. This corresponds to a fully inelastic impact. So far, the only index-3 schemes which can model elastic impacts within the discretization are the Paoli-Schatzman and the GGL scheme. Note in this context that the GGL scheme corresponds to an index-2 scheme with an additional projection.

A second way to incorporate impacts is to detect and evaluate these events apart from the time-stepping discretization. Two approaches exist: Either the impact positions and pre-impact velocities are obtained from interpolation, and an instantaneous impact problem is solved to obtain the new initial states for the further integration, see Anitescu et al. A second possibility is to reject a time step which contains an impact. Such a time step is recalculated using a Moreau like discrete set-valued law which includes also to the partially elastic impact case, see Stiegelmeyr and Funk. Impact treatment via event detection brings a lot of administrative effort, as separate routines for non-impulsive motion and impact treatment must be provided. Also the interpolation of an impact event might cause severe problems as the interpolation must reflect the mechanical behaviour exactly. In addition, such an approach can not handle accumulative switching points.

Let us finally give some future ideas for formulations on displacement level which incorporate partially elastic impact behaviour. As already mentioned, different choices for the position update formulas may cause impact like behaviour. So far only fully inelastic impact behaviour can modeled. Nevertheless, one could think of modifying the method of Jean. Instead of $\gamma_E^{s+1} = 0$ one could request $\gamma_E^{s+1} = -\varepsilon \gamma_E^s$ for $g_D^s + \gamma_E^s \Delta t = 0$ and $g_D^{s+1} + \gamma_E^{s+1} \Delta t = 0$. The problem might be to accomplish that the unilateral contacts remains closed for the two time steps s and $s+1$ in case of an impact. In addition, such unilateral contacts would not show stable behaviour in case of sustaining contact. However, if only impacts and no sustaining contact is expected, the method might work. Perhaps the most promising approach might be to incorporate a discrete Poisson impact law into a time-stepping scheme. The main problem is to guarantee that the discrete Poisson impact law holds for both impact and non-impulsive motion. Consider a point mass on a table. A decreasing external force acts on the point mass in such a way that the unilateral contact force between point mass and table decreases also, i.e. $\hat{P}_i^{s+1} < \hat{P}_i^s$. Consider now a Poisson based discrete set-valued law with $\varepsilon_p = 1$. This law would request that $\hat{P}_i^{s+1} \geq \varepsilon_p \hat{P}_i^s = \hat{P}_i^s$, which yields a contradiction. In order to overcome this contradiction, an idea for a time-stepping algorithm could be as follows: Calculate step s with $-g_{iE}^s \in \mathscr{N}_{\mathbb{R}_0^+}(\hat{P}_i^s)$ and step $s+1$ with $-g_{iE}^{s+1} \in \mathscr{N}_{\mathbb{R}_0^+}(\hat{P}_i^{s+1} - \varepsilon_p \hat{P}_i^s)$, where $\mathbf{h}$ must remain constant, i.e. $\mathbf{h}^s = \mathbf{h}^{s+1}$. An impact would then be treated between step s and step $s+1$, non-impulsive motion would be considered between the step $s+1$ and $s+2$.

Formulations on Velocity Level

Discrete set-valued laws on velocity level can be combined with impact laws of Newton type, which enables both elastic and inelastic impacts. In addition, such formulations allow for a unified treatment of geometric and kinematic set-valued laws. Furthermore, an active discrete geometric set-valued law $-\gamma_i \in \mathcal{N}_{\mathcal{C}_i}(\lambda_i)$ in constraint mode enforces the constraint equation $-\gamma_i = \mathbf{W}_i\mathbf{u} + \zeta_i = 0$, which is linear with respect to $\mathbf{u}$. In addition, discrete geometric set-valued laws which are non-active, i.e. which are in impressed mode, have not to be considered in the (expensive) evaluation of the discrete states $\hat{\sigma}$. The resulting index-2 time-stepping discretizations are simple in application and allow for a unified treatment of all kinds of set-valued laws. The most prominent index-2 time-stepping scheme is Moreau's midpoint rule, which has proven to be a simple and robust integrator for non-smooth systems. Other well-working algorithms come from Anitescu et al.

Unfortunately, formulations on velocity level might cause drift. These problems arise mostly for bilateral constraints with nonlinear gap functions $\mathbf{g}$. A unilateral contact may open from time to time which causes a possible drift to vanish. Drift problems occur mainly at unilateral contacts which remain closed for a longer time period, i.e. sustaining contacts. Such a sustaining contact is hardly possible for unilateral contacts with $\varepsilon = 1$. The drift problem of bilateral constraints might be tackled in the following ways: $\varepsilon \to 1$, additional projection to the admissible manifolds or drift stabilization. The first possibility can be considered for circle like trajectories, see also Sect. 5.1.2. The additional projection and the drift stabilization preserve the index-2 discretization and its advantages but formulate the constraints on displacement level. Note that the two approaches are also applicable for unilateral contacts, whereas the drift stabilization approach requests $\varepsilon = 0$. Unilateral contacts with ε close to one can not be stabilized, but the drift problem is also not essential for such contacts. Especially for bilateral constraints within a time-stepping scheme, the drift stabilization might be of interest. As the drift stabilization is purely explicit, it can be considered as an external loading in the vector $\mathbf{c}$. Therefore, it is no problem to combine drift stabilized and non-drift stabilized set-valued laws in the same algorithm, i.e. the unilateral contacts with $\varepsilon = 0$ and the bilateral constraints can be stabilized, the unilateral contacts with $\varepsilon > 0$ can be left on velocity level to enable partially elastic impacts.

5.3.9.4 Ill-Conditioning of the Time-Stepping Scheme

Index-3 schemes must use the unknowns $\Delta t\,\hat{\mathbf{P}}$ and $\Delta t\,\mathbf{u}$ to remain well-conditioned. Note that the ill-conditioning might not be obvious when iterating in $\hat{\mathbf{P}}$. When doing so, the influence of $\hat{\mathbf{P}}$ on the displacements is scaled by Δt, thus even large changes of $\hat{\mathbf{P}}^{\nu} \to \hat{\mathbf{P}}^{\nu+1}$ cause minor reaction $\mathbf{q}_E^{\nu} \to \mathbf{q}_E^{\nu+1}$. At a first glance, the scheme of Paoli and Schatzman is ill-conditioned due to the use of forces λ. However, the ill-conditioning is avoided by stating the problem solely in the displacements, i.e. the velocities and the forces λ which cause the ill-conditioning are eliminated. The approach is only applicable if the forces λ can be eliminated, which requests a regular Delassus matrix $\mathbf{G}$. In contrast to Paoli and Schatzman, Moreau uses percussions $\hat{\mathbf{P}}$ in his scheme which are well suited for impacts.

Note that within the discrete scheme it is possible to work with the forces λ instead of the percussions $\hat{\mathbf{P}}$. Such an approach is valid because a numerical scheme does not require impulsive forces and velocity jumps to satisfy the *discrete* set-valued laws, as these laws must only be enforced within a time step and not instantaneous. However, the resulting index-2 problem would be ill-conditioned. In this sense the ill-conditioning due to the use of λ is the "numerical consequence" of the instantaneous impact problem. Consider the following three cases:

- Non-smooth system without impact: The analytical description of such a system does only require finite forces λ. Note that such a system can always be expressed as an index-1 problem, of which the discretization remains well conditioned when using these forces λ.
- Non-smooth system with velocity jumps: The analytical solution requests impulsive forces in case of an impact, i.e. a percussion measure dP. Since velocity jumps can not be described on acceleration level, an index-1 representation of such a system is not possible. The index is at least two, and a discretization must use discrete percussions $\hat{\mathbf{P}}$ to remain well conditioned.
- Non-smooth systems with displacement jumps: Displacement jumps can not be enforced by impulsive forces within an analytical solution. Also velocities are not defined anymore. Systems with possible displacement jumps lead always to index-3 problems, which must be stated in the variables $\Delta t\,\mathbf{u}$ and $\hat{\mathbf{P}}\Delta t$ to remain well-conditioned.

Chapter 6
Augmented Time-Stepping by Step Size Adjustement and Extrapolation

This chapter investigates how the accuracy of Moreau's midpoint rule can be increased by step size adjustment and extrapolation. The idea is to perform time steps which contain switching points with a minimal step size Δt_{min}. Smooth time steps, i.e. time steps which do not contain switching points, are processed with larger steps sizes. Furthermore, extrapolation methods are used to increase the integration order in these time steps. The integration order of the classical Moreau midpoint rule is analyzed in Sect. 6.1. In Sect. 6.2 it is discussed how switching points can be localized, which yields a step size controlled time-stepping algorithm. Finally, in Sect. 6.3 extrapolation methods are applied to Moreau's midpoint rule, which requests some restrictions on these methods. Some examples are given in Sect. 6.5. The chapter is based on the original publications [92, 94].

6.1 Integration Order of Moreau's Midpoint Rule

This section aims at analyzing the integration order of Moreau's time-stepping scheme. Similar to the classical approach, the integration order is obtained by comparing the expansions of the exact and numeric solution with respect to the time t. The expansion of a non-smooth time step is obtained by combining the expansions of the associated smooth parts within the time step. Note that the sketched approach to obtain the integration order must be seen from an engineering point of view, i.e. it can not compete with much more sophisticated mathematical convergence proofs [28, 63].

6.1.1 Definition of the Integration Order

Consider a system of ordinary differential equations $\dot{\mathbf{y}} = \mathbf{f}(\mathbf{y},t)$ with initial conditions $\mathbf{y}(t_0) = \mathbf{y}_0$, of which an exact solution is $\mathbf{y}(t)$. Numerical schemes of the form

C. Studer: Numerics of Unilateral Contacts and Friction, LNACM 47, pp. 99–127.
springerlink.com

$$\mathbf{y}_{s+1} = \mathbf{y}_s + \Delta t\, \varphi(t_{s+1}, t_s, \mathbf{y}_s) \tag{6.1}$$

provide a sequence of values $\mathbf{y}_s$ which approximate the exact solution $\mathbf{y}(t_s)$ [86]. The function $\varphi(t_{s+1}, t_s, \mathbf{y}_s)$ is called incremental function of the integration scheme. The local integration error is defined as

$$\mathbf{e}_L(t_{s+1}) = \mathbf{y}(t_{s+1}) - \hat{\mathbf{y}}_{s+1} = \mathbf{y}(t_{s+1}) - \mathbf{y}(t_s) - \Delta t\, \varphi(t_{s+1}, t_s, \mathbf{y}(t_s)), \tag{6.2}$$

in which $\hat{\mathbf{y}}_{s+1}$ is the approximation obtained from one integration step starting from the exact value $\mathbf{y}(t_s)$. The terms $\mathbf{y}(t_{s+1})$ and $\hat{\mathbf{y}}_{s+1}$, which build the local integration error, can be written as a function of $\mathbf{y}(t_s)$ by using a Taylor-series expansion, provided that the trajectories are smooth. Comparing these expansions yields an estimation for the local integration error,

$$\max_{t_i} ||\mathbf{e}_L(t_i)|| \leq K_L \Delta t^{p+1} = \mathcal{O}(\Delta t^{p+1}), \tag{6.3}$$

in which K_L is an arbitrary bounded constant, which is independent of Δt. The scalar p is called the integration order. The global integration error gathers the accumulated local errors during the integration,

$$\mathbf{e}_G(t_{s+1}) = \mathbf{y}(t_{s+1}) - \mathbf{y}_{s+1} = \mathbf{y}(t_{s+1}) - \mathbf{y}_s - \Delta t\, \varphi(t_{s+1}, t_s, \mathbf{y}_s), \tag{6.4}$$

in which $\mathbf{y}_{s+1}$ is the approximation obtained after several integration steps. In the following, the global integration error $\mathbf{e}_G$ is expressed as sum of the local integration errors $\mathbf{e}_L$. It holds that

$$\begin{aligned} ||\mathbf{e}_G(t_{s+1})|| &= ||\mathbf{y}(t_{s+1}) - \mathbf{y}_{s+1}|| = ||\mathbf{y}(t_{s+1}) - \hat{\mathbf{y}}_{s+1} + \hat{\mathbf{y}}_{s+1} - \mathbf{y}_{s+1}|| \\ &= ||\mathbf{y}(t_{s+1}) - \hat{\mathbf{y}}_{s+1} + \mathbf{y}(t_s) - \mathbf{y}_s + \Delta t \Big(\varphi(t_{s+1}, t_s, \mathbf{y}(t_s)) - \varphi(t_{s+1}, t_s, \mathbf{y}_s) \Big) || \\ &\leq ||\mathbf{e}_L(t_{s+1})|| + (1 + L_\varphi \Delta t)\, ||\mathbf{e}_G(t_s)|| \\ &\leq ||\mathbf{e}_L(t_{s+1})|| + e^{L_\varphi \Delta t}\, ||\mathbf{e}_G(t_s)|| \end{aligned} \tag{6.5}$$

where L_φ is the Lipschitz constant of φ,

$$||\varphi(t_{s+1}, t_s, \mathbf{y}(t_s)) - \varphi(t_{s+1}, t_s, \mathbf{y}_s)|| \geq L_\varphi\, ||\mathbf{y}(t_s) - \mathbf{y}_s|| = L_\varphi\, ||\mathbf{e}_G(t_s)||. \tag{6.6}$$

The relation (6.5) allows for estimating the global error $\mathbf{e}_G(t_{s+1})$ by the global error $\mathbf{e}_G(t_s)$ and the local error $\mathbf{e}_L(t_{s+1})$. As a consequence, the global error $\mathbf{e}_G(t_{s+1})$ can also be estimated by a sum of local errors,

$$||\mathbf{e}_G(t_{s+1})|| \leq \sum_{i=1}^{s+1} \left(e^{L_\varphi\, (s+1-i)\, \Delta t}\, ||\mathbf{e}_L(t_i)|| \right) \leq \underbrace{e^{L_\varphi\, (t_{s+1} - t_0)}}_{:= K_p} \sum_{i=1}^{s+1} ||\mathbf{e}_L(t_i)||, \tag{6.7}$$

consider [86] for further explanations. The constant K_P in (6.7) describes the error propagation during the integration and depends only on the total integration time $t_{s+1} - t_0$. It characterizes the maximal amplification of an initial error during the integration process. There exist more sophisticated estimations for the global error, see for example [45, 86]. This work uses the conservative estimation (6.7) because it relates the sum of all local errors to the global error, which will be useful when dealing with different kinds of local errors. Combining the estimations (6.3) and (6.7) yields

$$||\mathbf{e}_G(t_{s+1})|| \leq K_P \sum_{i=1}^{s+1} ||\mathbf{e}_L(t_i)|| \leq K_P \frac{t_{s+1} - t_0}{\Delta t} K_L \Delta t^{p+1} = K_G \Delta t^p = \mathscr{O}(\Delta t^p), \quad (6.8)$$

in which $K_G = K_P K_L (t_{s+1} - t_0)$ is an arbitrary bounded constant which is independent from Δt. The term $\frac{t_{s+1}-t_0}{\Delta t}$ corresponds to the number of time steps s. Note that a local integration error $\mathbf{e}_L$ of order $p+1$ results in a global integration error $\mathbf{e}_G$ of order p. The corresponding integration scheme is called an order p integration method.

Example

Consider the method of Heun, which has the form

$$\mathbf{y}_{s+1} = \mathbf{y}_s + \frac{\Delta t}{2}\Big(\mathbf{f}(t_s,\mathbf{y}_s) + \mathbf{f}\big(t_{s+1},\mathbf{y}_s + \Delta t\,\mathbf{f}(t_s,\mathbf{y}_s)\big)\Big). \quad (6.9)$$

Expansion of the incremental function in t and $\mathbf{y}$ yields

$$\mathbf{y}_{s+1} = \mathbf{y}_s + \frac{\Delta t}{2}\left(\mathbf{f}(t_s,\mathbf{y}_s) + \mathbf{f}(t_s,\mathbf{y}_s) + \left.\frac{\partial \mathbf{f}}{\partial t}\right|_{t_s,\mathbf{y}_s} \Delta t + \left.\frac{\partial \mathbf{f}}{\partial \mathbf{y}}\right|_{t_s,\mathbf{y}_s} \Delta t\,\mathbf{f}(t_s,\mathbf{y}_s) + \mathscr{O}(\Delta t^2)\right). \quad (6.10)$$

The exact solution can be approximated by the Taylor-series

$$\mathbf{y}(t_{s+1}) = \mathbf{y}(t_s) + \dot{\mathbf{y}}(t_s)\Delta t + \ddot{\mathbf{y}}(t_s)\frac{\Delta t^2}{2} + \mathscr{O}(\Delta t^3), \quad (6.11)$$

with

$$\dot{\mathbf{y}}(t_s) = \mathbf{f}\big(t_s,\mathbf{y}(t_s)\big), \quad (6.12)$$

$$\ddot{\mathbf{y}}(t_s) = \left.\frac{\partial \mathbf{f}}{\partial t}\right|_{t_s,\mathbf{y}(t_s)} + \left.\frac{\partial \mathbf{f}}{\partial \mathbf{y}}\frac{\partial \mathbf{y}}{\partial t}\right|_{t_s,\mathbf{y}(t_s)} = \left.\frac{\partial \mathbf{f}}{\partial t}\right|_{t_s,\mathbf{y}(t_s)} + \left.\frac{\partial \mathbf{f}}{\partial \mathbf{y}}\right|_{t_s,\mathbf{y}(t_s)} \mathbf{f}\big(t_s,\mathbf{y}(t_s)\big). \quad (6.13)$$

Comparing (6.10) with (6.11) yields a local integration order of three. Note in this context that the approximation $\mathbf{y}_s$ in (6.10) is replaced by the exact value $\mathbf{y}(t_s)$ when calculating the local integration error $\mathbf{e}_L$. The global integration error of Heun's method is two.

6.1.2 *Expansion of the Exact Solution* $\mathbf{q}(t)$ *and* $\mathbf{u}(t)$

In this section we will derive Taylor-series expansions of the exact solution $\mathbf{q}(t)$ and $\mathbf{u}(t)$ for the cases smooth motion, non-impulsive and impulsive switching point and accumulation point. A non-impulsive switching point corresponds to a jump in the accelerations, an impulsive switching point to a jump in the velocities. In Sect. 6.1.4, these expansions will be compared with the expansion of the time-stepping scheme in order to estimate the local integration error.

6.1.2.1 Smooth Motion

Consider the index-1 DAE

$$\dot{\mathbf{q}} = \mathbf{u}, \tag{6.14}$$

$$\dot{\mathbf{u}} = \mathbf{M}^{-1}\Big((\mathbf{W}+\mathbf{Q})\boldsymbol{\lambda}+\mathbf{h}\Big), \tag{6.15}$$

$$\dot{\gamma} = \mathbf{W}^{\top}\dot{\mathbf{u}}+\hat{\zeta} = 0, \tag{6.16}$$

which describes the motion of a mechanical system with respect to a constant state σ_1. Set-valued force laws which are in constraint mode are considered by the generalized force directions $\mathbf{W} = \mathbf{W}(\sigma_1, \mathbf{q}, t)$ and by the forces $\boldsymbol{\lambda}$. Set-valued force laws in impressed mode are considered by the vector of all external and gyroscopic forces $\mathbf{h} = \mathbf{h}(\sigma_1, \mathbf{q}, \mathbf{u}, t)$ and by the matrix $\mathbf{Q} = \mathbf{Q}(\sigma_1, \mathbf{q}, t)$, which is required for sliding Coulomb friction elements. Using the kinematic relation (6.16), the index-1 DAE (6.14-6.16) is reformulated to become a system of first order ordinary differential equations,

$$\dot{\mathbf{q}} = \mathbf{u}, \tag{6.17}$$

$$\dot{\mathbf{u}} = {}_1\mathbf{k}, \tag{6.18}$$

in which ${}_1\mathbf{k} = \mathbf{k}(\sigma_1, \mathbf{q}, \mathbf{u}, t)$ and

$$\mathbf{k} = \mathbf{k}(\sigma, \mathbf{q}, \mathbf{u}, t) = \mathbf{M}^{-1}\Big[\mathbf{h} - \Big(\mathbf{W}+\mathbf{Q}\Big)\Big(\mathbf{W}^{\top}\mathbf{M}^{-1}(\mathbf{W}+\mathbf{Q})\Big)^{-1}\Big(\mathbf{W}^{\top}\mathbf{M}^{-1}\mathbf{h}+\hat{\zeta}\Big)\Big]. \tag{6.19}$$

From the ODE (6.17-6.18) one obtains the Taylor-expansion for a smooth time step $\Delta t = t_E - t_B$ with constant state σ_1,

$$\mathbf{q}_E = \mathbf{q}_B + \dot{\mathbf{q}}\big|_B \Delta t + \ddot{\mathbf{q}}\big|_B \Delta t^2 + \mathscr{O}(\Delta t^3) = \mathbf{q}_B + \mathbf{u}_B \Delta t + {}_1\mathbf{k}_B \frac{\Delta t^2}{2} + \mathscr{O}(\Delta t^3), \tag{6.20}$$

$$\mathbf{u}_E = \mathbf{u}_B + \dot{\mathbf{u}}\big|_B \Delta t + \mathscr{O}(\Delta t^2) = \mathbf{u}_B + {}_1\mathbf{k}_B \Delta t + \mathscr{O}(\Delta t^2), \tag{6.21}$$

in which ${}_1\mathbf{k}_B = \mathbf{k}(\sigma_1, \mathbf{q}_B, \mathbf{u}_B, t_B)$.

6.1.2.2 Switching Point

Assume that the time step $\Delta t = t_E - t_B$ contains a non-impulsive or impulsive switching point at time t_S, where $t_B < t_S < t_E$. Firstly, we divide such a time step into two smooth substeps without any specification on the switching point. Secondly, we distinguish between non-impulsive switching points for which the velocities do not jump, i.e. $\mathbf{u}_S^+ = \mathbf{u}_S^-$, and impulsive switching points for which the velocities jump, i.e. $\mathbf{u}_S^+ \neq \mathbf{u}_S^-$.

In the first part of the time step $\Delta t_1 = t_S - t_B$, the state is σ_1, and the motion is described by an underlying differential equation system

$$\dot{\mathbf{q}} = \mathbf{u}, \tag{6.22}$$

$$\dot{\mathbf{u}} = {}_1\mathbf{k}. \tag{6.23}$$

At the time t_S, a new state σ_2 is determined to obtain a new underlying differential equation system

$$\dot{\mathbf{q}} = \mathbf{u}, \tag{6.24}$$

$$\dot{\mathbf{u}} = {}_2\mathbf{k}, \tag{6.25}$$

associated with the second part of the time step $\Delta t_2 = t_E - t_S$. Note that ${}_1\mathbf{k}$ and ${}_2\mathbf{k}$ are completely different functions which correspond to the different states σ_1 and σ_2, respectively. It does not make sense to incorporate ${}_1\mathbf{k}$ and ${}_2\mathbf{k}$ into the expansion of the exact solution, because the two terms have to be considered with different step sizes Δt_1 and Δt_2, which do not appear in the time-stepping method. The expansion for the first part of the time step $\Delta t_1 = t_S - t_B$ yields

$$\mathbf{q}_S = \mathbf{q}_B + \mathbf{u}_B \Delta t_1 + \mathcal{O}(\Delta t_1^2), \tag{6.26}$$

$$\mathbf{u}_S^- = \mathbf{u}_B + \mathcal{O}(\Delta t_1). \tag{6.27}$$

The expansion for the second part of the time step $\Delta t_2 = t_E - t_S$ becomes

$$\mathbf{q}_E = \mathbf{q}_S + \mathbf{u}_S^+ \Delta t_2 + \mathcal{O}(\Delta t_2^2), \tag{6.28}$$

$$\mathbf{u}_E = \mathbf{u}_S^+ + \mathcal{O}(\Delta t_2). \tag{6.29}$$

Note that ${}_1\mathbf{k}$ and ${}_2\mathbf{k}$ are included by the terms $\mathcal{O}(\Delta t_1)$, $\mathcal{O}(\Delta t_1^2)$, $\mathcal{O}(\Delta t_2^2)$ and $\mathcal{O}(\Delta t_2)$.

Non-impulsive Switching Point

If the switching point is not impulsive, then the pre- and the post-impact velocity $\mathbf{u}_S^-$ and $\mathbf{u}_S^+$ are identical. Inserting the expansions (6.26) and (6.27) into the expansions (6.28) and (6.29) yields

$$\mathbf{q}_E = \mathbf{q}_B + \mathbf{u}_B \Delta t + \mathcal{O}(\Delta t^2), \tag{6.30}$$
$$\mathbf{u}_E = \mathbf{u}_B + \mathcal{O}(\Delta t). \tag{6.31}$$

Impulsive Switching Point

At an impulsive event (impact), the pre- and post-impact velocity $\mathbf{u}_S^-$ and $\mathbf{u}_S^+$ at the time t_S are not the same anymore. In this case, the impact equations and Newton's impact law must be applied,

$$0 = \mathbf{M}_S(\mathbf{u}_S^+ - \mathbf{u}_S^-) - (\mathbf{W}_S + \mathbf{Q}_S)\Lambda, \tag{6.32}$$
$$0 = \gamma_S^+ + \varepsilon \gamma_S^-, \tag{6.33}$$
$$\gamma_S^\pm = \mathbf{W}_S \mathbf{u}_S^\pm + \zeta_S, \tag{6.34}$$

where the impact state is assumed to be σ_3. The impulsive forces and the generalized force directions of the set-valued laws which participate in the impact are denoted by Λ and $\mathbf{W}_S = \mathbf{W}(\sigma, \mathbf{q}_S, t)$, respectively. The matrix $\mathbf{Q}_S = \mathbf{Q}(\sigma_3, \mathbf{q}_S, t)$ considers sliding Coulomb friction elements. The equations (6.32-6.34) are solved for the post-impact velocities $\mathbf{u}_S^+$,

$$\mathbf{u}_S^+ = \mathbf{u}_S^- + {}_3\mathbf{K}_S\, {}_3\gamma_S^-, \tag{6.35}$$

where ${}_3\mathbf{K}_S = \mathbf{K}(\sigma_3, \mathbf{q}_S, t_S)$. Note that both ${}_3\gamma_S^-$ and ${}_3\mathbf{K}_S$ depend on the state σ_3. The matrix $\mathbf{K}$ itself is

$$\mathbf{K} = \mathbf{K}(\sigma, \mathbf{q}, t) = -\mathbf{M}^{-1}\Big(\mathbf{W} + \mathbf{Q}\Big)\Big(\mathbf{W}^\top \mathbf{M}^{-1}(\mathbf{W} + \mathbf{Q})\Big)^{-1}\Big(\mathbf{I} + \varepsilon\Big). \tag{6.36}$$

The matrix $\mathbf{K}$ does not depend on the velocities $\mathbf{u}$ and is therefore continuous at the time instance t_S. Equation (6.35) can now be used to link equations (6.26-6.27) with equations (6.28-6.29), which yields

$$\mathbf{q}_E = \mathbf{q}_B + \mathbf{u}_B \Delta t + {}_3\mathbf{K}_S\, {}_3\gamma_S^- \Delta t_2 + \mathcal{O}(\Delta t^2), \tag{6.37}$$
$$\mathbf{u}_E = \mathbf{u}_B + {}_3\mathbf{K}_S\, {}_3\gamma_S^- + \mathcal{O}(\Delta t). \tag{6.38}$$

Finally, the term ${}_3\mathbf{K}_S\, {}_3\gamma_S^-$ can be approximated by

$${}_3\mathbf{K}_S\, {}_3\gamma_S{}^- = {}_3\mathbf{K}_B\, {}_3\gamma_B + \mathcal{O}(\Delta t_1), \tag{6.39}$$

which yields for the expansion in the case of an impact,

$$\mathbf{q}_E = \mathbf{q}_B + \mathbf{u}_B \Delta t + {}_3\mathbf{K}_B\, {}_3\gamma_B \Delta t_2 + \mathcal{O}(\Delta t^2), \tag{6.40}$$
$$\mathbf{u}_E = \mathbf{u}_B + {}_3\mathbf{K}_B\, {}_3\gamma_B + \mathcal{O}(\Delta t). \tag{6.41}$$

6.1.2.3 Accumulation Points

Consider the case of time steps with more than one switching point. If these switching points are not accumulative, then the step size can be refined to obtain only time steps with one switching point, which have already been discussed. In the case of accumulative switching points, the time span between the switching points tends towards zero, and even refining the step size will not lead to time steps with only a single switching point. Let the impact times be denoted by t_{S_i}, $i = 1 \dots n$. The decomposition of the time step in smooth parts then yields by (6.26-6.29)

$$\begin{array}{lll}
\mathbf{q}_{S_1} = \mathbf{q}_B + \mathbf{u}_B \Delta t_0 + \mathcal{O}(\Delta t_0^2) & \mathbf{u}_{S_1}^- = \mathbf{u}_B + \mathcal{O}(\Delta t_0) & \mathbf{u}_{S_1}^+ = \mathbf{u}_{S_1}^- + {}_1\mathbf{K}_{S_1}\, {}_1\gamma_{S_1}^- \\
\vdots & \vdots & \vdots \\
\mathbf{q}_{S_i} = \mathbf{q}_{S_{i-1}} + \mathbf{u}_{S_{i-1}} \Delta t_{i-1} + \mathcal{O}(\Delta t_{i-1}^2) & \mathbf{u}_{S_i}^- = \mathbf{u}_{S_{i-1}} + \mathcal{O}(\Delta t_{i-1}) & \mathbf{u}_{S_i}^+ = \mathbf{u}_{S_i}^- + {}_i\mathbf{K}_{S_i}\, {}_i\gamma_{S_i}^- \\
\vdots & \vdots & \\
\mathbf{q}_E = \mathbf{q}_{S_n} + \mathbf{u}_{S_n} \Delta t_n + \mathcal{O}(\Delta t_n^2) & \mathbf{u}_E = \mathbf{u}_{S_n} + \mathcal{O}(\Delta t_n) &
\end{array}$$

The presence of an accumulative switching point requires very small relative velocities ${}_i\gamma_{S_i}$, because otherwise the time span between the switching points would not tend towards zero. If the relative velocities ${}_i\gamma_{S_i} \approx 0$ are neglected, then the expansion for the accumulative switching point becomes

$$\mathbf{q}_E = \mathbf{q}_B + \mathbf{u}_B \Delta t + \mathcal{O}(\Delta t^2), \tag{6.42}$$

$$\mathbf{u}_E = \mathbf{u}_B + \mathcal{O}(\Delta t). \tag{6.43}$$

In this sense the accumulation point is looked upon as non-impulsive switching point, i.e. the velocity jumps in the accumulation point are neglected.

6.1.3 Expansion of Moreau's Time-Stepping Scheme

In this section we will derive a Taylor-series expansion of the time-stepping scheme of Moreau. In Sect. 6.1.4, this expansion will be compared to the expansions of the exact solution $\mathbf{q}(t)$ and $\mathbf{u}(t)$ obtained in Sect. 6.1.2 in order to estimate the local integration error.

Moreau's time-stepping scheme for a discrete state $\hat{\sigma}_4$ yields

$$\mathbf{q}_E = \mathbf{q}_B + \frac{\mathbf{u}_E + \mathbf{u}_B}{2} \Delta t, \tag{6.44}$$

$$\mathbf{u}_E = \mathbf{u}_B + \mathbf{M}_M^{-1} \Big((\mathbf{W}_M + \mathbf{Q}_M) \hat{\mathbf{P}} + \mathbf{h}_M \Delta t \Big), \tag{6.45}$$

$$0 = \gamma_E + \varepsilon \gamma_B = \mathbf{W}_M^\top \mathbf{u}_E + \zeta_E + \varepsilon (\mathbf{W}_M^\top \mathbf{u}_B + \zeta_B), \tag{6.46}$$

All discrete set-valued laws in constraint mode are considered by the discrete percussions $\hat{\mathbf{P}}$ and by the generalized force directions $\mathbf{W}_M = \mathbf{W}(\hat{\sigma}_4, \mathbf{q}_M, t_M)$. Discrete set-valued laws in impressed mode are considered by the vector of external and gyroscopic forces $\mathbf{h}_M = \mathbf{h}(\hat{\sigma}_4, \mathbf{q}_M, \mathbf{u}_B, t_M)$ or by the matrix $\mathbf{Q}_M = \mathbf{W}(\hat{\sigma}_4, \mathbf{q}_M, t_M)$. Elimination of the discrete percussions $\hat{\mathbf{P}}$ yields

$$\mathbf{q}_E = \mathbf{q}_B + \frac{\mathbf{u}_E + \mathbf{u}_B}{2}\Delta t, \tag{6.47}$$

$$\mathbf{u}_E = \mathbf{u}_B + {}_4\hat{\mathbf{k}}_M \Delta t + {}_4\mathbf{K}_M\, {}_4\gamma_B, \tag{6.48}$$

in which ${}_4\hat{\mathbf{k}}_M = \hat{\mathbf{k}}(\hat{\sigma}_4, \mathbf{q}_M, \mathbf{u}_B, t_M)$, ${}_4\mathbf{K}_M = \mathbf{K}_M(\hat{\sigma}_4, \mathbf{q}_M, t_M)$ according to (6.36) and

$$\hat{\mathbf{k}} = \mathbf{M}^{-1}\Big[\mathbf{h} - \Big(\mathbf{W}+\mathbf{Q}\Big)\Big(\mathbf{W}^\top \mathbf{M}^{-1}(\mathbf{W}+\mathbf{Q})\Big)^{-1}\Big(\mathbf{W}^\top \mathbf{M}^{-1}\mathbf{h} + \frac{\zeta_E - \zeta_B}{\Delta t}\Big)\Big]. \tag{6.49}$$

Expanding the terms ${}_4\hat{\mathbf{k}}_M$ and ${}_4\mathbf{K}_M\, {}_4\gamma_B$ yields

$${}_4\hat{\mathbf{k}}_M = {}_4\hat{\mathbf{k}}_B + \mathcal{O}(\Delta t), \tag{6.50}$$

and

$${}_4\mathbf{K}_M\, {}_4\gamma_B = \Big({}_4\mathbf{K}_B + \mathcal{O}(\Delta t)\Big)\, {}_4\gamma_B. \tag{6.51}$$

The expansion of the time-stepping scheme therefore becomes

$$\mathbf{q}_E = \mathbf{q}_B + \mathbf{u}_B\Delta t + \Big({}_4\mathbf{K}_B + \mathcal{O}(\Delta t)\Big)\, {}_4\gamma_B \frac{\Delta t}{2} + {}_4\hat{\mathbf{k}}_B \frac{\Delta t^2}{2} + \mathcal{O}(\Delta t^3), \tag{6.52}$$

$$\mathbf{u}_E = \mathbf{u}_B + \Big({}_4\mathbf{K}_B + \mathcal{O}(\Delta t)\Big)\, {}_4\gamma_B + {}_4\hat{\mathbf{k}}_B \Delta t + \mathcal{O}(\Delta t^2). \tag{6.53}$$

Consider now the terms ${}_4\hat{\mathbf{k}}_B$ and ${}_4\mathbf{k}_B$ from (6.49) and (6.19), respectively. Note that the terms do only distinguish by $\frac{\zeta_E - \zeta_B}{\Delta t}$ in (6.49) and $\hat{\zeta}_B$ in (6.19). Note that for a time step without a switching point the relation

$$\begin{aligned}\frac{\zeta_E - \zeta_B}{\Delta t} &= \frac{1}{\Delta t}\Big(\gamma_E - \gamma_B - \mathbf{W}_M^\top(\mathbf{u}_E - \mathbf{u}_B)\Big) = \\ &= \frac{1}{\Delta t}\Big((\dot{\gamma}_B - \mathbf{W}_M^\top \dot{\mathbf{u}}_B)\Delta t + \mathcal{O}(\Delta t^2)\Big) = \hat{\zeta}_B + \mathcal{O}(\Delta t),\end{aligned} \tag{6.54}$$

holds, where we have taken into account that

$$\gamma_E = \gamma_B + \dot{\gamma}_B \Delta t + \mathcal{O}(\Delta t^2), \tag{6.55}$$

$$\mathbf{u}_E = \mathbf{u}_B + \dot{\mathbf{u}}_B \Delta t + \mathcal{O}(\Delta t^2). \tag{6.56}$$

By using relation (6.54), we obtain

$${}_4\hat{\mathbf{k}}_B = {}_4\mathbf{k}_B + \mathcal{O}(\Delta t), \tag{6.57}$$

which causes the expansion of a smooth time step to become

$$\mathbf{q}_E = \mathbf{q}_B + \mathbf{u}_B \Delta t + \left({}_4\mathbf{K}_B + \mathcal{O}(\Delta t)\right) {}_4\gamma_B \frac{\Delta t}{2} + {}_4\mathbf{k}_B \frac{\Delta t^2}{2} + \mathcal{O}(\Delta t^3), \tag{6.58}$$

$$\mathbf{u}_E = \mathbf{u}_B + \left({}_4\mathbf{K}_B + \mathcal{O}(\Delta t)\right) {}_4\gamma_B + {}_4\mathbf{k}_B \Delta t + \mathcal{O}(\Delta t^2). \tag{6.59}$$

In the case of a time step which contains a switching point, the expansion

$$\mathbf{q}_E = \mathbf{q}_B + \mathbf{u}_B \Delta t + {}_4\mathbf{K}_B\, {}_4\gamma_B \frac{\Delta t}{2} + \mathcal{O}(\Delta t^2), \tag{6.60}$$

$$\mathbf{u}_E = \mathbf{u}_B + {}_4\mathbf{K}_B\, {}_4\gamma_B + \mathcal{O}(\Delta t), \tag{6.61}$$

applies, where ${}_4\mathbf{k}_B$ is included in $\mathcal{O}(\Delta t)$ and $\mathcal{O}(\Delta t^2)$.

6.1.4 Local Integration Order

Comparing the expansion of the exact solution with the expansion of the time-stepping scheme yields the local integration errors. We distinguish between the following cases: a smooth time step, a time step with one non-impulsive switching point and a time step with one impact. Furthermore, some comments concerning accumulative switching points are given.

6.1.4.1 Time Step Without Switching Point (Smooth Motion)

Comparing the expansion (6.20-6.21) with the expansion (6.58-6.59) yields the local integration error for time steps without switching points. It is assumed that the state σ_1 within the smooth time step is equal to the discrete state $\hat{\sigma}_4$, which yields $\mathbf{k}_{1,B} = \mathbf{k}_{4,B}$. In the case of smooth motion, the relative velocities ${}_4\gamma_B$ are very small, as otherwise the discrete set-valued laws would not remain in constraint mode, i.e. a unilateral contact would open or an impact would occur. Neglecting the relative velocities ${}_4\gamma_B$ yields a local integration order of three for the displacements $\mathbf{q}$ and of two for the velocities $\mathbf{u}$,

$$\max_{t_i} ||\mathbf{e}_{Lq}(t_i)|| \leq K_{Lq} \Delta t^3 = \mathcal{O}(\Delta t^3), \tag{6.62}$$

$$\max_{t_i} ||\mathbf{e}_{Lu}(t_i)|| \leq K_{Lu} \Delta t^2 = \mathcal{O}(\Delta t^2). \tag{6.63}$$

6.1.4.2 Time Step with Non-Impulsive Switching Point (Jump Accelerations)

The expansion (6.30-6.31) and the expansion (6.60-6.61) yield the local integration error for time steps with one non-impulsive switching point. It still holds that ${}_4\gamma_B \approx$

0, as else an impact would occur. Neglecting this term yields a local integration order of two for the displacements $\mathbf{q}$ and of one for the velocities $\mathbf{u}$,

$$\max_{t_i} ||\mathbf{e}_{Lq}|| \leq K_{Lq}\Delta t^2 = \mathscr{O}(\Delta t^2), \tag{6.64}$$

$$\max_{t_i} ||\mathbf{e}_{Lu}|| \leq K_{Lu}\Delta t = \mathscr{O}(\Delta t). \tag{6.65}$$

6.1.4.3 Time Step with Impulsive Switching Point (Jump Velocities)

Comparing the expansion (6.40-6.41) with the expansion (6.60-6.61) yields the local integration error for a time step with one impact. It is assumed that the discrete state $\hat{\sigma}_4$ of the time-stepping scheme is equal to the state σ_3 of the impact law. In this case, it holds that ${}_4\mathbf{K}_B = {}_3\mathbf{K}_B$ and ${}_4\gamma_B = {}_3\gamma_B$, which yields a local integration order of one for both the displacements $\mathbf{q}$ and the velocities $\mathbf{u}$,

$$\max_{t_i} ||\mathbf{e}_{Lq}|| \leq K_{Lq}\Delta t = \mathscr{O}(\Delta t), \tag{6.66}$$

$$\max_{t_i} ||\mathbf{e}_{Lu}|| \leq K_{Lu}\Delta t = \mathscr{O}(\Delta t). \tag{6.67}$$

6.1.4.4 Time Step with Accumulative Switching Point

Subtracting the expansion (6.42-6.43) from the expansion (6.60-6.61) yields the local integration error for a time step with an accumulative switching point. In the case of an accumulative switching point, the relative velocity ${}_4\gamma_B$ is very small and therefore expected to be negligible. Under this assumption, the local integration error for the displacements $\mathbf{q}$ and velocities $\mathbf{u}$ is

$$\max_{t_i} ||\mathbf{e}_{Lq}|| \leq K_{Lq}\Delta t = \mathscr{O}(\Delta t^2), \tag{6.68}$$

$$\max_{t_i} ||\mathbf{e}_{Lu}|| \leq K_{Lu}\Delta t = \mathscr{O}(\Delta t). \tag{6.69}$$

6.1.5 Global Integration Order

Equation (6.7) is used to relate the local errors of the single time steps (6.62-6.69) to a global integration error of the time-stepping scheme. All smooth time steps are assigned to an index set $\mathscr{I}_1$ and all switching time steps to an index set $\mathscr{I}_2$. This distinction between the local errors of smooth time steps and the local errors of switching time steps allows for writing the global error as

$$||\mathbf{e}_G(t_{s+1})|| \leq K_P\Big(\sum_{\substack{i=1\\ i\in\mathscr{I}_1}}^{s+1} ||\mathbf{e}_L(t_i)|| + \sum_{\substack{i=1\\ i\in\mathscr{I}_2}}^{s+1} ||\mathbf{e}_L(t_i)||\Big). \tag{6.70}$$

The first sum in (6.70) adds all local errors of the smooth time steps, the second sum adds all local errors of of the switching time steps. We conservatively estimate the local integration error of the state $(\mathbf{q}, \mathbf{u})$ of smooth and switching time steps by

$$\text{smooth time step}: \max_{t_i} ||\mathbf{e}_{L_{q,u}}|| \leq K_L \Delta t^2, \tag{6.71}$$

$$\text{switching time step}: \max_{t_i} ||\mathbf{e}_{L_{q,u}}|| \leq K_L \Delta t. \tag{6.72}$$

It is important to realize that a change in the step size Δt has mainly an influence on the number of smooth time steps. The number of switching time steps, i.e. time steps which contain a switching point, an impact or an accumulative switching point, depends mainly on the system evolution and not on the step size. Let n_2 be the number of switching time steps. In this case, the number of smooth time steps is

$$n_1 = \frac{t_{s+1} - t_0}{\Delta t} - n_2, \tag{6.73}$$

and the global integration error becomes by (6.8)

$$||\mathbf{e}_G(t_{s+1})|| \leq K_P \Big(\Big(\frac{t_{s+1} - t_0}{\Delta t} - n_2 \Big) K_L \Delta t^2 + n_2 K_L \Delta t \Big) \leq K_G \Delta t. \tag{6.74}$$

Thus, Moreau's time-stepping scheme is expected to be an integration scheme of order one. Most important to realize is the fact that the local integration errors of switching time steps add to the global integration error with almost no reduction of the exponent of Δt, because the number of switching time steps does not mainly depend on the chosen step size. Of course there exist exceptions, e.g. in the field of granular media. For such systems, every time step is a switching time step and the step size Δt defines the number of switching points which are provided to resolve the systems evolution. As a consequence, one can not speak of an order-one integration method with respect to the movement of a single particle. When simulating granular media with Moreau's time-stepping method, not the movement of a single particle but the overall behaviour of the media is of interest. This overall behaviour does not seem to react very sensitive on the chosen step size. However, a mathematical proof for this claim is missing.

6.2 Step Size Adjustment for Switching Points

This section deals with an easy to manage and uniquely working switching point detection. As already discussed, a switching point occurs when a set-valued law switches from impressed to constrained mode or vice versa. In Sect. 5.2, a state σ has been introduced, and a switching time instance t_S has been defined as $\sigma^-(t_S) \neq \sigma^+(t_S)$. This state σ refers to the exact time evolution of the system, i.e. $\sigma(t)$ is the actual state of the system at time t. The discrete state $\hat{\sigma}$, which has been introduced in Sect. 5.3, is associated with a time-stepping scheme. It characterizes the state on which the discretization of the time step is based. While σ applies for a single time instance, $\hat{\sigma}$ is valid for a whole time step. Only if the step size becomes small, the evolution of σ and $\hat{\sigma}$ can become similar.

6.2.1 Determining the Discrete State $\hat{\sigma}$

Consider Moreau's time-stepping scheme formulated as projective equations (5.54). The system is solved by iteration in $\hat{\mathbf{P}}$, for example by a JORprox or SORprox method. The behaviour of the projective equations (5.54) is now investigated at the solution $\hat{\mathbf{P}}^{\infty}$ of such an iteration, i.e. $\hat{\mathbf{P}}^{\nu} \to \hat{\mathbf{P}}^{\infty}$. Either the $\text{prox}_{\hat{\mathscr{A}}_i}$ has no action, i.e.

$$\hat{\mathbf{P}}_i^{\infty} = \text{prox}_{\hat{\mathscr{A}}_i}\big(\hat{\mathbf{P}}_i^{\infty} - r_i(\sum_{j=1}^{n} \mathbf{G}_{ij}\hat{\mathbf{P}}_j^{\infty} + \mathbf{c}_i)\big) = \hat{\mathbf{P}}_i^{\infty} - r_i(\sum_{j=1}^{n} \mathbf{G}_{ij}\hat{\mathbf{P}}_j^{\infty} + \mathbf{c}_i), \tag{6.75}$$

which yields

$$\sum_{j=1}^{n} \mathbf{G}_{ij}\hat{\mathbf{P}}_j^{\infty} + \mathbf{c}_i = \gamma_{Ei} + \varepsilon_i \gamma_{Bi} = 0, \quad \hat{\mathbf{P}}_i^{\infty} \in \hat{\mathscr{A}}_i, \tag{6.76}$$

or the $\text{prox}_{\hat{\mathscr{A}}_i}$ really projects on the set

$$\hat{\mathbf{P}}_i^{\infty} = \text{prox}_{\hat{\mathscr{A}}_i}\big(\hat{\mathbf{P}}_i^{\infty} - r_i(\sum_{j=1}^{n} \mathbf{G}_{ij}\hat{\mathbf{P}}_j^{\infty} + \mathbf{c}_i)\big) = \mathbf{k}_i. \tag{6.77}$$

In the first case, $\hat{\mathbf{P}}_i$ lies inside the admissible set $\hat{\mathscr{A}}_i$ to enforce the relation $\gamma_{Ei} + \varepsilon_i \gamma_{Bi} = 0$, which yields the discrete set-valued law i to be in constrained mode within the entire time step. The discrete state is assigned to be $\hat{\sigma}_i = 0$. In the second case, the $\text{prox}_{\hat{\mathscr{A}}_i}$ function impresses a value $\mathbf{k}_i$, and the discrete set-valued law i is assumed to be in impressed mode. The discrete state is then $\hat{\sigma}_i = 1$. Thus, by observing the projection behaviour of the $\text{prox}_{\hat{\mathscr{A}}_i}$ function in the solution point $\hat{\mathbf{P}}^{\infty}$, it can easily be decided whether the discrete set-valued laws are in constrained or in impressed mode. This detection procedure is independent of any physical behaviour, i.e. it only looks for a switch in the mathematical structure of the discretization. Therefore, the approach is very robust and consistent with the algorithm, because it evaluates what the discretization is doing and not what the discretization is expected to do, i.e. it respects the decisions made by the time-stepping algorithm and does not consult any further physical criterion to evaluate which discrete state $\hat{\sigma}$ is applicable. The detection procedure is applicable to all kinds of discrete set-valued laws, which keeps the implementation easy and small.

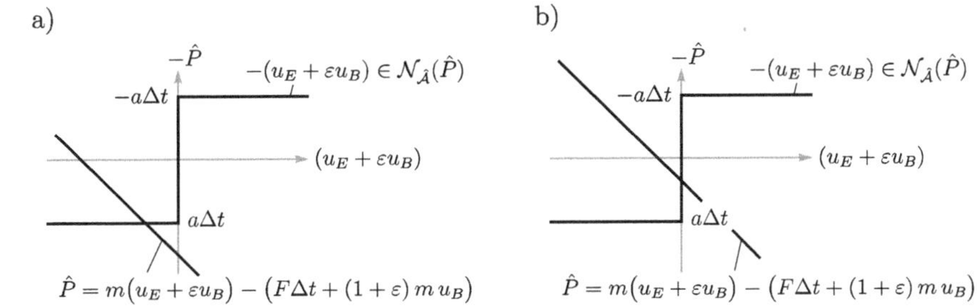

Fig. 6.1 Graphical interpretation of (6.78). Figure **a** depicts the sliding, Fig. **b** the sticking case

Example

Consider a point mass, which can move along a table in x direction. Between the point mass and table acts friction with a maximal friction force $a=\mu mg$. In addition, a force F pulls the point mass in x direction. The velocity in x direction is denoted by u. The discrete time-stepping scheme for this example reads

$$m(u_E-u_B)-F\Delta t-\hat{P}=0,\quad -(u_E+\varepsilon u_B)\in\mathcal{N}_{\hat{\mathscr{A}}}(\hat{P}),\quad \hat{\mathscr{A}}=[-a\Delta t;a\Delta t]. \tag{6.78}$$

Figure 6.1 provides a graphical interpretation of (6.78). Elimination of the velocity u_E yields an inclusion expressed solely in terms of the discrete percussion $\hat{P}$,

$$-\Big(\frac{F\Delta t}{m}+\frac{\hat{P}}{m}+(1+\varepsilon)u_B\Big)\in\mathcal{N}_{\hat{\mathscr{A}}}(\hat{P}). \tag{6.79}$$

The associated projective equation becomes

$$\hat{P}=\operatorname{prox}_{\hat{\mathscr{A}}}\Big(\hat{P}-r\big(\frac{F\Delta t}{m}+\frac{\hat{P}}{m}+(1+\varepsilon)u_B\big)\Big). \tag{6.80}$$

It is assumed that the velocity u_B at the beginning of the time step is zero. The behaviour of the system is now investigated for different values of the external force F. If F is smaller than the maximal friction force a, that means $|F|<a$, then the point mass will stick on the table and the iterative solution of the projective equation (6.80) has a fixed point at $\hat{P}=-F\Delta t$,

$$-F\Delta t=\operatorname{prox}_{\hat{\mathscr{A}}}\big(-F\Delta t\big), \tag{6.81}$$

in which the $\operatorname{prox}_{\hat{\mathscr{A}}}$ does not change the value of its argument and $\hat{\sigma}$ is set to $\hat{\sigma}=0$. The discrete percussion fulfills the constraint $u_E=u_B=0$, which can easily be verified by inserting $\hat{P}=-F\Delta t$ in the equations of motion (6.78). If the external force F is larger than the maximal friction force a, i.e. $F>a$, then the point mass will slide in positive direction on the table and the iterative solution of the projective equation (6.80) has a fixed point at $\hat{P}=-a\Delta t$,

$$-a\,\Delta t = \mathrm{prox}_{\hat{\mathscr{A}}}\big(\underbrace{-a\,\Delta t - r\frac{\Delta t}{m}(F-a)}_{<-a\,\Delta t}\big), \tag{6.82}$$

in which the $\mathrm{prox}_{\hat{\mathscr{A}}}$ projects on the boundary of the set $\hat{\mathscr{A}} = [-a\,\Delta t, a\,\Delta t]$ and impresses a discrete percussion $\hat{P} = -a\,\Delta t$, for which we set $\hat{\sigma} = 1$.

6.2.2 *Localizing Switching Points*

A change in the discrete state $\hat{\sigma}$ indicates a switching point. Consider a non-smooth system with four set-valued laws, for which the discrete states $\hat{\sigma}$ obtained by time-stepping are depicted in Fig. 6.2. We recognize that the first discrete set-valued law switches its discrete state $\hat{\sigma}_1$ from 0 in the second time step to 1 in the third time step. Thus, the corresponding switching point will be located between t_1 and t_3. Note that the changed discrete state $\hat{\sigma}_1$ in the third time step does not necessarily imply that the switching point is located in this time step, as it is also possible that it is located somewhere at the end of the second time step: A time step $t_B^s \to t_E^s$ containing a switching point at t_S, i.e. $t_B^s \leq t_S \leq t_E^s$, has either the discrete state $\hat{\sigma}^s = \sigma^-(t_S)$ or $\hat{\sigma}^s = \sigma(t_S^+)$. The first case yields $\hat{\sigma}^s \neq \hat{\sigma}^{s+1}$, the second $\hat{\sigma}^s \neq \hat{\sigma}^{s-1}$. If a change $\hat{\sigma}^s \neq \hat{\sigma}^{s+1}$ in the discrete state occurs, then the switching point t_S can be located either in the time step s or $s+1$.

Examples

Consider a unilateral contact, which is open at the midpoint $\mathbf{q}_M$, but closed by the end t_E of the time step. In this case, the discrete state is $\hat{\sigma} = 1$, and $\sigma(t_E) = 0$. Another example is given by the friction element discussed in the example on page 111. Assume that the discussed point mass is sliding on the table having a velocity u_B, and assume that no external force F is present. Due to the sliding force $-a$ the point mass will stop at $t_S = t_B + \frac{u_B m}{a}$, i.e. $\sigma(t < t_S) = 1$ and $\sigma(t > t_S) = 0$. Consider the corresponding projective equation (6.80),

$$\hat{P} = \mathrm{prox}_{\hat{\mathscr{A}}}\Big(\hat{P} - r\big(\frac{\hat{P}}{m} + (1+\varepsilon)u_B\big)\Big), \tag{6.83}$$

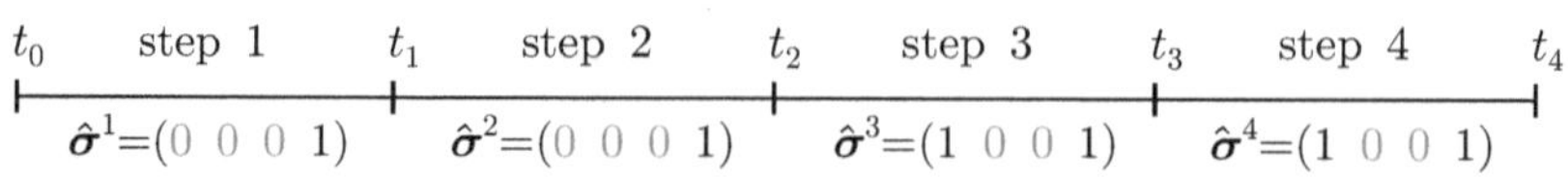

Fig. 6.2 Time evolution of a non-smooth system with four set-valued laws. Below the time step the discrete state $\hat{\sigma}$ of the four discrete set-valued laws is depicted. The third time step has a discrete state $\hat{\sigma}$ different to its predecessor, and the associated switching point will be located between t_1 and t_3

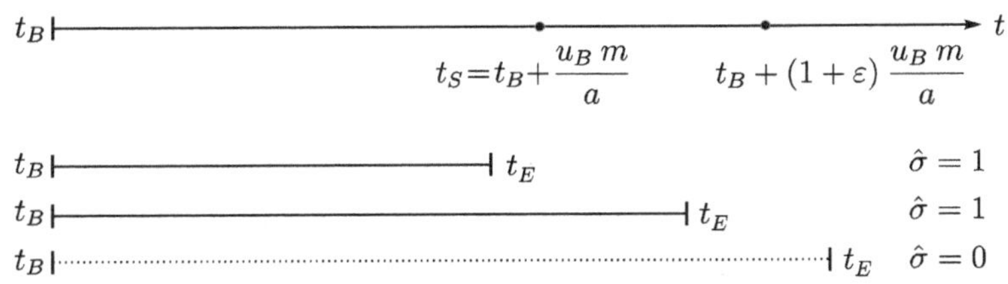

Fig. 6.3 Visualization of the switching point detection. A change of the discrete state $\hat{\sigma}$ is indicated by a change of the line style (*solid to dotted*). The first time step does not contain the switching point and has a discrete state $\hat{\sigma} = 1$. The second and the third depicted time step contain the switching point. In contrast to the second time step, the third time step detects the switching point, i.e. its discrete state changes to $\hat{\sigma} = 0$. The discrete scheme switches its discrete state at $\Delta t = (1+\varepsilon)\frac{u_B m}{a}$, which contradicts the "true" location of the switching point

in which ε denotes the tangential restitution coefficient. The $\text{prox}_{\mathscr{A}}$ will only project to $\hat{P} = -a\Delta t$ if

$$-a\Delta t > \hat{P} - r\Big(\frac{\hat{P}}{m} + (1+\varepsilon)u_B\Big) = -a\Delta t - r\Big(\frac{-a\Delta t}{m} + (1+\varepsilon)u_B\Big), \tag{6.84}$$

which can be simplified to

$$\Delta t < (1+\varepsilon)\frac{u_B m}{a}. \tag{6.85}$$

Consider the time step $t_B \rightarrow t_E$ which contains the slip-stick transition at t_S, i.e. $t_B < t_S < t_E$, $\sigma(t_B) = \sigma(t_S^-) = 1$ and $\sigma(t_S^+) = \sigma(t_E) = 0$. If $t_S < t_E < t_B + (1+\varepsilon)u_B\frac{m}{a}$, then the $\text{prox}_{\mathscr{A}}$ still projects and the discrete state will be $\hat{\sigma} = 1 \neq \sigma(t_S^+)$. The switching point is only detected within this time step if $t_S < t_B + (1+\varepsilon)u_B\frac{m}{a} < t_E$, which yields the $\text{prox}_{\mathscr{A}}$ to be inactive. In case of $\varepsilon = 0$ only the latter case applies and the switching point is always detected by its hosting time step. See also Fig. 6.3 for a graphical visualization. Note that the switching point of the discrete scheme at $\Delta t = (1+\varepsilon)\frac{u_B m}{a}$ contradicts the "true" location of the switching point. The condition for sliding within the discrete scheme, i.e. $u_E + \varepsilon u_B > 0$, $\hat{P} = -a\Delta t$ is not violated as long as (6.85) holds. This can also easily be verified by consulting the associated equations of motion (6.78).

A switching point can be located by a regula falsi method. In case of a changing discrete state $\hat{\sigma}$ the algorithm drops back to the beginning of the predecessor time step and continues the integration with a smaller step size until another change in the discrete state $\hat{\sigma}$ is detected. Again, the algorithm drops back and repeats this procedure until a minimal step size Δt is reached which resolves the switching point properly. After having resolved the switching point, the step size is again increased until a maximal user defined step size Δt_{max} is reached. This procedure is depicted in Fig. 6.4. Note that in the case of a changing discrete state $\hat{\sigma}$, the algorithm drops back to the beginning of the predecessor time step, because the switching point can be located either in the predecessor or in the actual time step. Both, the actual and

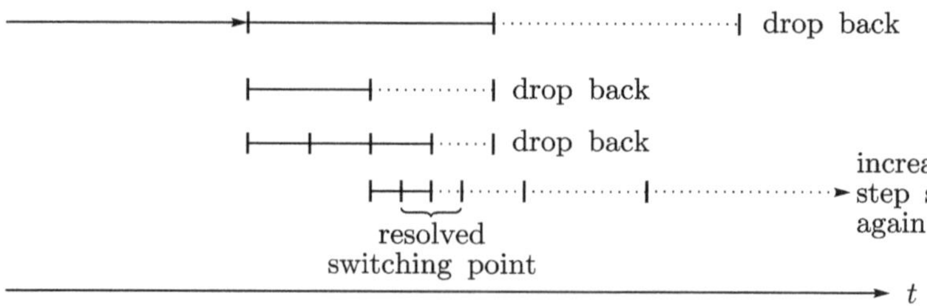

Fig. 6.4 Switching point search by a regula falsi method. A change of the discrete state $\hat{\sigma}$ is indicated by a change of the line style (*solid to dotted*) [92, 94]

the predecessor time step are rejected. A more detailed algorithmic description of this procedure will be given in Sect. 6.4.

Remark

Note the differences between the switching point localization by regula falsi and by root finding, see Figs. 6.4 and 5.14. While the root finding approach is intended to find the exact switching time, i.e. a single time instance, the regula falsi approach aims only at localizing switching points more precisely in order to reduce the step size. Considering accumulative switching points, the exact switching time localization by root finding will cause the algorithm to fail. Problems may also occur if $\varepsilon \neq 0$, which might lead to situations for which the $\text{prox}_{\mathscr{C}}$ function remains in the old discrete state, i.e. $\hat{\sigma} = \sigma^-(t_S)$, see also Fig. 6.3. The combination of root finding and extrapolation might also be problematic, see 6.3.1. Note in this context that the time-stepping algorithm yields an order one (or two in the case of Potra) approximation. The transition time Δt_1 is based on this order one (or order two) approximation, which is different from the higher order approximation constructed by the extrapolation procedure. These different approximations might contradict each other, which would cause the exact switching point localization to fail. Note that these problems do not occur for the regula falsi approach as this approach is not intended to localize the switching points exactly. The regula falsi approach aims only at reducing the step sizes near switching points.

6.3 Higher Order Integration for Smooth Time Steps

As mentioned in Sect. 5.2, the time evolution of non-smooth systems can be divided into smooth parts and switching points, or, in terms of the time-stepping algorithm, in smooth time steps $t_B \rightarrow t_E$ with constant state $\sigma(t), t_B \leq t \leq t_E$, and in switching intervals. In this section it is explained how the integration order of smooth Moreau time steps can be increased. It does not make any sense to apply

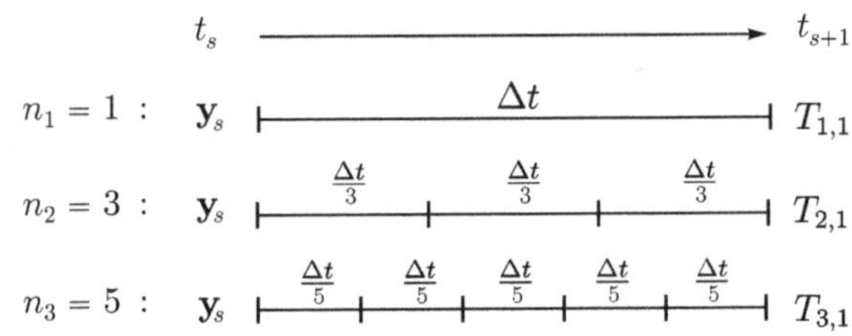

Fig. 6.5 Different approximations for $T_{i,1}$.

higher order integration to switching intervals as the time evolution of the states $\mathbf{q}$ and $\mathbf{u}$ within this interval is not smooth and does not allow for an asymptotic expansion of the integration error. In Sect. 6.3.1 the classical extrapolation methods are shortly reviewed. Detailed information about these integration schemes can be found in [26, 45, 58, 86, 94]. In Sect. 6.3.2, extrapolation is applied to Moreau's time-stepping scheme in order to improve the integration order.

6.3.1 Extrapolation

Consider an arbitrary base integration scheme of order p. Extrapolation methods allow for constructing an order $(p+k-1)$ approximation using k approximations calculated with a base integration scheme. In Fig. 6.5, a typical integration procedure is depicted. The value $\mathbf{y}_s$ is the approximation of $\mathbf{y}(t)$ at t_s. Starting from $\mathbf{y}_s$ we calculate with the base integration scheme different approximations $T_{i,1}$ for $\mathbf{y}(t_{s+1})$ by dividing the time interval $\Delta t = t_{s+1} - t_s$ into a different number n_i of substeps with step size $\Delta t/n_i$. The index i represents the index of the approximation, the index 1 indicates that the approximation has the same integration order as the base integration scheme. For example, a first approximation $T_{1,1}$ is obtained by performing $n_1 = 1$ step Δt with the base integration scheme. A second approximation $T_{2,1}$ is calculated by $n_2 = 3$ substeps $\Delta t/3$, a third approximation $T_{3,1}$ by $n_3 = 5$ substeps $\Delta t/5$. Note that also other choices of the number of substeps may be used, for example $n_1 = 1, n_2 = 2, n_3 = 3$.

By combination of the different approximations $T_{i,1}$, approximations $T_{i,j}$ can be obtained which have a higher integration order than the original base integration scheme. The index j indicates the increase of order. This process of increasing the integration order by combination of the approximations $T_{i,1}$ is referred to as extrapolation methods. The methods are based on the asymptotic expansion of the global integration error and are therefore only applicable for smooth motion.

The values $T_{i,j}$ are obtained by the Aitken Neville formula

$$T_{i,j+1} = T_{i,j} + \frac{T_{i,j} - T_{i-1,j}}{\frac{n_i}{n_{i-j}} - 1}, \tag{6.86}$$

and are stored in an so-called extrapolation tableau of the form

$$\begin{array}{llll} T_{1,1} & & & \\ T_{2,1} & T_{2,2} & & \\ T_{3,1} & T_{3,2} & T_{3,3} & \\ \vdots & \vdots & \vdots & \ddots \end{array} \tag{6.87}$$

The extrapolation tableau (6.87) is built up in the following way: The approximations $T_{1,1}$ and $T_{2,1}$ are calculated in order to obtain $T_{2,2}$. The approximations $T_{2,1}$ and $T_{3,1}$ give $T_{3,2}$, which results together with $T_{2,2}$ in the value $T_{3,3}$, and so on. The approximations $T_{2,2}$ and $T_{3,2}$ yield an order increase of one, the approximation $T_{3,3}$ yields an order increase of two compared with $T_{i,1}$. The extrapolation tableau is successively built up until the required accuracy $|T_{i+1,i+1} - T_{i,i}| < tol$ is reached, or until a predefined integration order is achieved.

Example

This small example aims at the numerical solution of the differential equation $y'(x) = 3x^2$ with $y(0) = 0$, of which the exact solution is $y(x) = x^3$. As base scheme the explicit Euler rule is taken, $y_{i+1} = y_i + y'(x_i)\Delta x$, which becomes in this case $y_{i+1} = y_i + 3x_i^2 \Delta x$. In order to obtain $y(1)$, one time step with $\Delta x = 1$, three time steps with $\Delta x = 1/3$ and five time steps with $\Delta x = 1/5$ are performed. The extrapolation tableau yields

$$\begin{array}{lll} T_{1,1} = 0 & & \\ T_{2,1} = \frac{5}{9} & T_{2,2} = \frac{5}{9} + \frac{\frac{5}{9}-0}{\frac{3}{1}-1} = \frac{5}{6} & \\ T_{3,1} = \frac{18}{25} & T_{3,2} = \frac{18}{25} + \frac{\frac{18}{25}-\frac{5}{9}}{\frac{5}{3}-1} = \frac{29}{30} & T_{3,3} = \frac{29}{30} + \frac{\frac{29}{30}-\frac{5}{6}}{\frac{5}{1}-1} = 1 \end{array} \tag{6.88}$$

The third order approximation $T_{3,3} = 1$ gives the exact result $y(1) = 1$.

6.3.2 Extrapolation Applied to Moreau's Midpoint Rule

Moreau's time-stepping scheme is now taken as base integration scheme for extrapolation. Note that this approach makes only sense for smooth time steps. The application of extrapolation or any other higher order integration method on non-smooth switching intervals contradicts the asymptotic expansion of the integration error.

In smooth parts of the motion, different approximations $T_{i,1}$ are gained by dividing the main time step in different substeps, which are processed by the time-stepping method. Based on these approximations $T_{i,1}$, higher order approximations can be computed using extrapolation. Note that all integration steps within a smooth main time step have to be performed on the same discrete state $\hat{\sigma}$. If the discrete state $\hat{\sigma}$ changes, then a switching point is present and the extrapolation process must be aborted.

Opposed to the classical extrapolation, only an uneven number of substeps n_i, i.e. $n_1 = 1$, $n_2 = 3$, $n_3 = 5$ and so on should be used. Even numbers of substeps, for example $n_2 = 2$, lead to numerical problems, as discussed in detail in [94]. The reason for this failure is the representation of the discrete set-valued laws in constraint mode, i.e.

$$-(\gamma_{Ei} + \varepsilon_i \gamma_{Bi}) = 0 \quad \Rightarrow \quad \gamma_{Ei} = -\varepsilon_i \gamma_{Bi}, \quad \varepsilon_i > 0, \tag{6.89}$$

which is not really a valid integration scheme but rather a mapping. We recognize that the step size Δt does not occur in (6.89). Doing $n_1 = 1$ substep Δt with (6.89) yields $T_{1,1} = -\varepsilon_i \gamma_{Bi}$, i.e. $T_{1,1}$ and γ_{Bi} have *different* signs. Performing $n_2 = 2$ substeps $\Delta t/2$ with (6.89) results in $T_{2,1} = (-\varepsilon_i)^2 \gamma_{Bi}$, i.e. $T_{2,1}$ and γ_{Bi} have the *same* sign. Thus, if we calculate different approximations $T_{i,1}$ with (6.89) using a mix of even and uneven numbers of substeps, then we obtain different signs for $T_{i,1}$. In [94] it is shown that this sign change causes the extrapolation process to fail. As the case $n_1 = 1$ should always be included due to the resolution of impulsive switching points, we must restrict ourselves to a uneven number of substeps to guarantee the same sign for all the approximations $T_{i,1}$.

6.4 Overall Algorithm

In Sect. 6.2 step size adjustment has been used to localize switching points. Time steps which contain such switching incidents have been resolved by a minimal step size Δt_{min}. In Sect. 6.3 extrapolation has been used to increase the integration order of smooth Moreau time steps with step size Δt_{max}. Note that extrapolation is applied on both, velocities $\mathbf{u}$ and the displacements $\mathbf{q}$. This section combines the two ideas, i.e. step size adjustment and extrapolation, to obtain an overall augmented time-stepping algorithm. By doing so, it is possible to increase the global integration order with respect to Δt_{max}. This section analyzes first the overall integration order and gives a short overview on a possible structure of the algorithm [94].

6.4.1 Integration Order

Consider equation (6.70), which writes the global error e_G as a sum of the local errors e_L. It is distinguished between local errors of smooth and switching time steps. The smooth time steps are gathered in the index set $\mathscr{I}_1$ and the switching time steps in the index set $\mathscr{I}_2$. When using step size adjustment and extrapolation, the local errors become

$$\text{smooth time step}: \max_{t_i} ||\mathbf{e}_L|| < K_L \Delta t_{max}^p, \tag{6.90}$$

$$\text{switching time step}: \max_{t_i} ||\mathbf{e}_L|| < K_L \Delta t_{min}. \tag{6.91}$$

According to Sect. 6.1.5, the global error is

$$||\mathbf{e}_G(t_{m+1})|| \leq K_P\Big(\Big(\frac{t_{m+1}-t_0}{\Delta t} - n_2\Big) K_L \Delta t_{max}^p + n_2 K_L \Delta t_{min}\Big). \tag{6.92}$$

Next, Δt_{min} is written as $K_t \Delta t_{max}^p$ with K_t being a constant. Doing so, equation (6.92) can be expressed solely in Δt_{max}, i.e.

$$||\mathbf{e}_G(t_{m+1})|| \leq K_G \Delta t_{max}^p, \tag{6.93}$$

Thus the global error is of order p with respect to Δt_{max}. The relation $\Delta t_{min} = K_t \Delta t_{max}^p$ must be seen in the following context: if the step size Δt_{max} is reduced, i.e. $\Delta t_{max} \to \frac{\Delta t_{max}}{n}$, then the associated reduction of the minimal step size is $\Delta t_{min} \to \frac{\Delta t_{min}}{n^p}$.

Note that the relation $\Delta t_{min} = K_t \Delta t_{max}^p$ does not specify the choice of the minimal step size Δt_{min}, as K_t is free of choice. In general the minimal step size should be chosen small. However, if the minimal step size Δt_{min} is too small, i.e. around the working precision of the processor, rounding errors destroy the integration order of the method. A maximal step size Δt_{max} which is too large may cause problems in the detection of closed unilateral contacts.

It is questionable to speak of an overall integration order p in the classical sense, because still integration steps are used with a local integration order of one for switching points. Only the piecewise smooth parts of the motion are approximated by an order p scheme. If the maximal step size Δt_{max} is used to estimate the local integration errors of switching intervals, then a pseudo higher order estimation with respect to Δt_{max} is obtained. To conclude, an overall integration order of p with respect to Δt_{max} is achieved by restricting the local error of switching intervals by a small minimal step size $\Delta t_{min} = K_t \Delta t_{max}^p$ and by restricting the local error of smooth time steps by a higher integration order p in the classical sense.

Remark

Note that until now it was assumed that switches are instantaneous actions. However, one could also accept that switches include a switching time span, which corresponds then to the minimal step size Δt_{min}. In this sense, it is possible to use also a minimal step size Δt_{min} which is independent from Δt_{max}, i.e. a switching time span Δt_{min} which is given by the model. The estimation (6.92) is then written without the term $n_2 K_L \Delta t_{min}$.

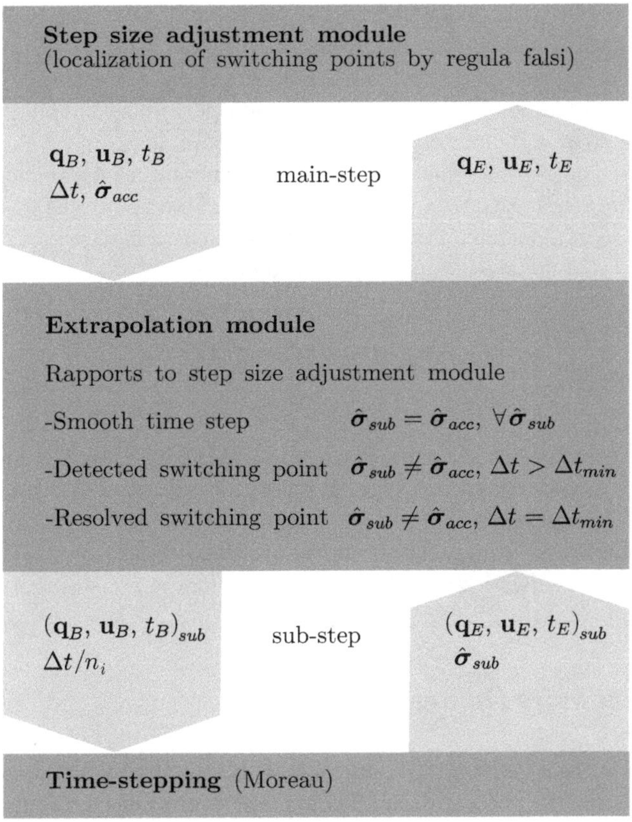

Fig. 6.6 Overall algorithm with step size adjustment, extrapolation and time-stepping module. The step size adjustment module passes the position $\mathbf{q}_B$, the velocity $\mathbf{u}_B$ and the time t_B to the extrapolation module. In addition, an accepted discrete state $\hat{\sigma}_{acc}$ is provided which is currently applicable. The extrapolation module calculates the different approximations $T_{i,1}$ by calling the time-stepping module. It monitors the discrete state $\hat{\sigma}_{sub}$ of the different substeps $\Delta t/n_i$ in order to detect switching points. It rapports to the step size adjustment module whether all substeps have been performed on $\hat{\sigma}_{acc}$ (smooth step), or whether a switching point was detected, i.e. $\hat{\sigma}_{acc} \neq \hat{\sigma}_{sub}$. If the step size Δt is equal to Δt_{min}, then a detected switching point is assumed to be resolved.

6.4.2 Implementation

Figure 6.6 depicts a possible structure of an overall algorithm, which consists of three modules: the step size adjustment module, the extrapolation module and the time-stepping module. The step size adjustment module controls the simulation and strings together the different time steps. The module steers the step size and holds information about the discrete state $\hat{\sigma}_{acc}$ which is currently applicable. In order to perform a main time step, the step size adjustment module calls the extrapolation module. This module splits the main time step into substeps and calls the time stepping module in order to obtain the different approximations $T_{i,1}$. The time-stepping module reports the discrete state $\hat{\sigma}$ of the set-valued force laws to the extrapolation module. This enables the latter module to detect switching points, i.e. if all approximations $T_{i,1}$ are based on the same accepted discrete state $\hat{\sigma}_{acc}$, the extrapolation module assumes a smooth time-step and performs extrapolation, else, the extrapolation process is aborted and a switching point is assumed. A report is sent to the step-size adjustment module, which either continues the integration in case of smooth time steps or starts a switching point localization based on a regula falsi method in case of a detected switching point. A detailed discussion concerning the implementation of the algorithm can be found in [94].

6.5 Examples

This section presents some examples of non-smooth systems for which the time evolution has been obtained by using the algorithm described above [92, 94]. First, three simple examples are presented, i.e. a point mass falling on a table, a point mass sliding on a table and a single DOF impact oscillator. Next, the woodpecker toy is discussed, which is a well-known benchmark problem in non-smooth mechanics.

6.5.1 Point Mass Falling on a Table

Consider a point mass which falls on a table. When the point mass hits the table, an impact occurs and the point mass is reflected. The restitution coefficient is chosen as $\varepsilon = 0.7$, which causes the point mass to jump less high after each impact. The time lapse between the impacts tends versus zero and ends up in an accumulation point, i.e. infinitely many impacts in a finite time. After the accumulation point, the point mass will remain on the table. The following parameters have been used: initial height $z_0 = 0.07$ m, mass $m = 1$ kg, minimal step size $\Delta t_{min} = 10^{-5}$s and maximal step size $\Delta t_{max} = 0.05$s. In Fig. 6.7 the time evolution of the step size Δt and of the integration order as well as the time evolution of the height z and of its derivative $\dot{z}$ are shown. Note that the step size increases after the simulation has passed the accumulation point. In the smooth part of the motion the integration order is two which yields the exact result for the parabolic path of $z(t)$.

6.5.2 Point Mass Sliding on a Table

Consider a point mass which slides on a table, see Fig. 6.8a. The following parameters apply: mass $m = 1$ kg, maximal friction force $a = \mu m g = 2$ N, tangential restitution coefficient $\varepsilon = 0$, initial velocity $\mathbf{u}_0 = (2.0, -0.2)^\top$ m/s. In addition, an external force $F = (-2.5, -0.25)^\top$ N acts on the point mass during the first three

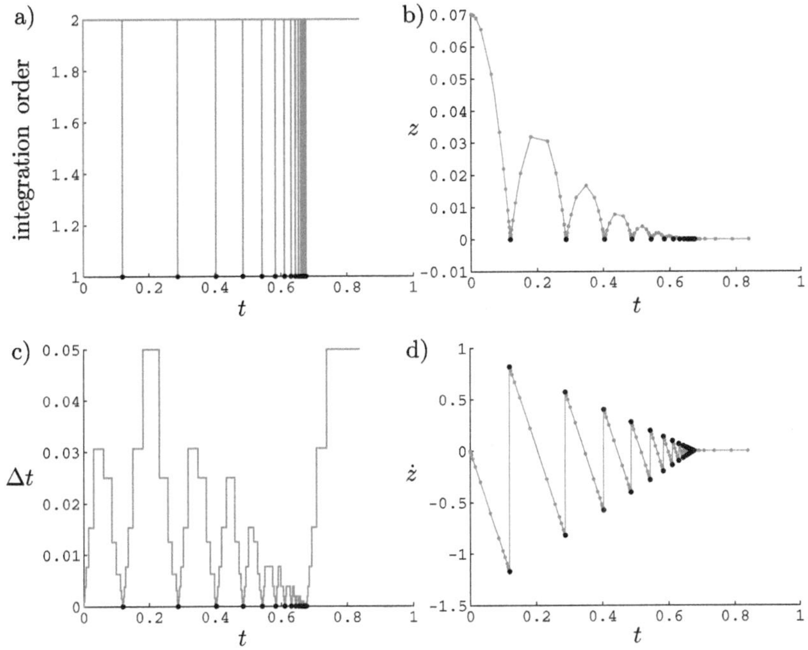

Fig. 6.7 Simulation results of a point mass falling on a table. In Figs. **a-d**, the time evolution of the step size Δt, the integration order, the height z and its derivative $\dot{z}$ are shown [92, 94]

seconds. A plot of the path in the x-y-plane is depicted in Fig. 6.8b). The time evolutions of the integration order, the step size Δt, the friction force λ and of the absolute velocity of the point mass are depicted in Figs. 6.8c-f. Due to the external force F the point mass will change its sliding direction almost to the opposite direction. The absolute velocity decreases, and a small step size together with a high integration order are used in this part of the simulation. Note that the sliding force in the x-direction changes its direction, and that the sliding force in y-direction has a peak at the turning point. The absolute value of the sliding force remains 2N. After the point mass has turned its direction, it is accelerated in the direction of the external force. The displacements are described by a parabolic polynomial, thus an integration order of two is sufficient. After three seconds the external force vanishes, and the friction force decelerates the point mass until it sticks.

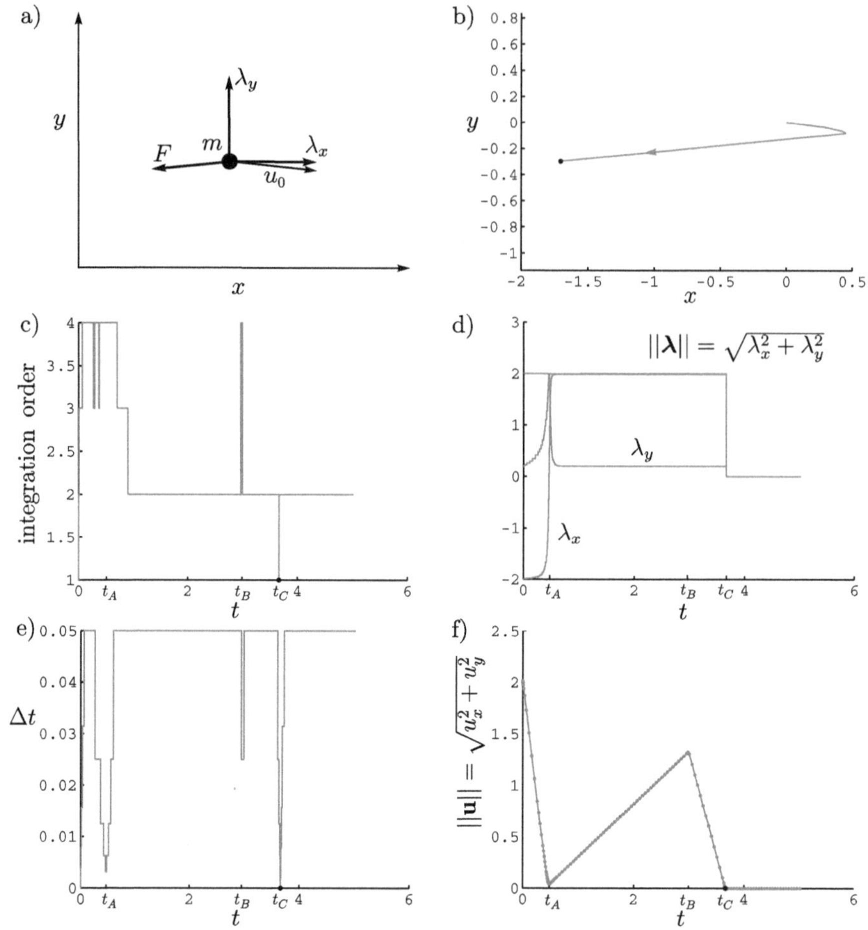

Fig. 6.8 Point mass sliding on a table. Figure **b** shows the path in the x-y-plane. Furthermore, the time evolutions of the integration order, the step size Δt, the friction force λ and the absolute velocity of the point mass are depicted in Figs. **c**-**f**. At the time instances t_A, t_B and t_C the point mass turns around, the force is switched off and a slip-stick transition occurs, respectively [94]

6.5.3 Single DOF Impact Oscillator

Consider a single DOF impact oscillator which consists of a mass, a spring and a unilateral contact which disturbs the oscillation of the mass. The position of the point mass is addressed by the coordinate x, the velocity by its derivative $\dot{x}$. The following parameters have been used: mass $m = 0.1$ kg, spring stiffness $c = 20$ N/m, restitution coefficient $\varepsilon = 0.6$, initial height $x_0 = -0.5$ m and initial velocity $\dot{x} = 0.2$ m/s. The spring is unstressed for $x = -0.15$ m. The minimal step size was chosen $\Delta t_{min} = K_t \Delta t_{max}^p$ with $K_t = 1\,\mathrm{s}^{p-1}$. Figures 6.9b-c depict the position x and the velocity $\dot{x}$ of the point mass.

In contrast to the other examples, the time evolution of this example has been obtained by using a time step adjustment module which uses a fixed integration order in the smooth time intervals, i.e. the integration order is *not* determined by an error estimation $T_{n,n} - T_{n-1,n-1}$ but is given by the user. The example aims at investigating the relation between the global error e_G, the integration order p, the maximal steps size Δt_{max} and the average step size $\Delta t_{average}$, see Fig. 6.9d and e. The average step size is obtained by dividing the simulation time by the number of required steps. Note that the maximal step size and the average step size do not depend linearly as for larger maximal step sizes the effort to come down to the minimal step size increases. A double-logarithmic plot of the global error e_G versus the maximal step size Δt_{max} should be more or less a straight line with slope $s = p$ if the global error e_G is approximatively equal to $K_G \Delta t_{max}^p$. See also Sect. 6.4.1 for further explanatory notes. Considering the double-logarithmic plot of the global error e_G versus the average step size $\Delta t_{average}$, the slopes s become larger than the p, which can be explained by assuming a relation of the kind $\Delta t_{max} = \Delta t_{average}^{\alpha}$ with $\alpha > 1$.

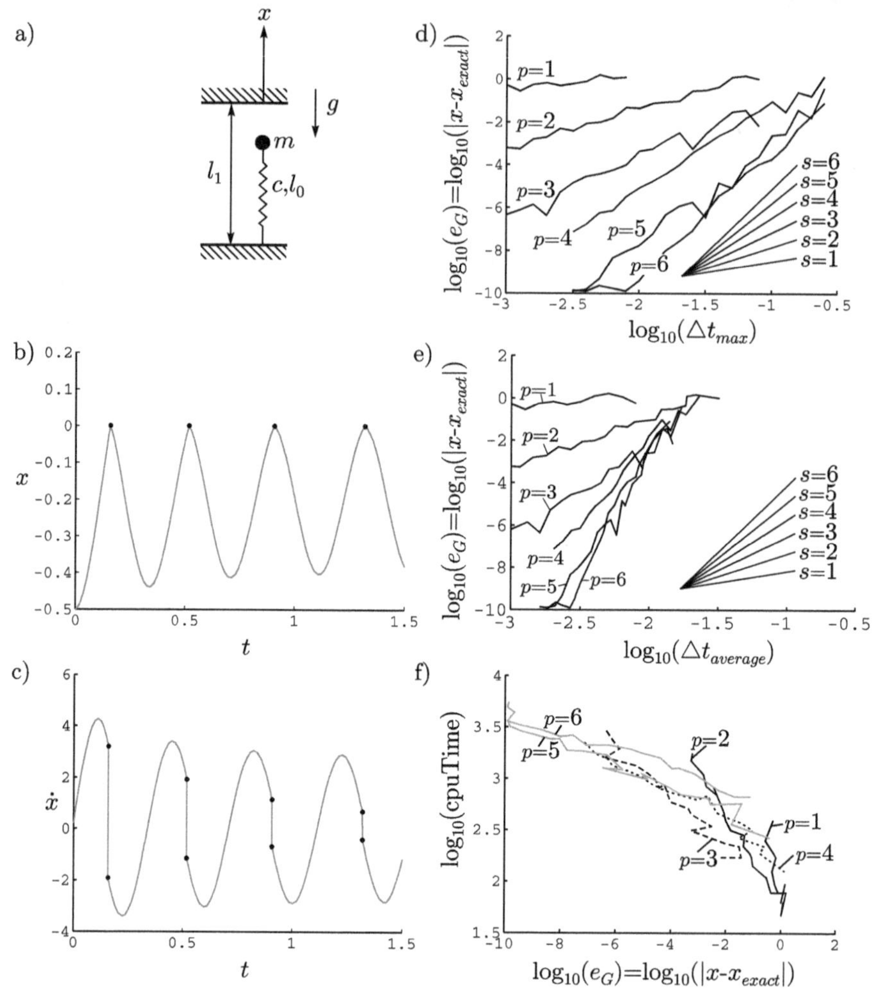

Fig. 6.9 Single DOF impact oscillator. Figure **b** and **c** show the position x and the velocity $\dot{x}$ of the mass. In Figs. **d** and **e**, the global error $e_G = |x - x_{exact}|$ is depicted as a function of the maximal step size Δt_{max} and the average step size $\Delta t_{average}$ for different calculations with integration order p. Plotting the global error versus the maximal steps size yields approximately straight lines with slope $s = p$. Using the average step size results in approximately straight lines with slope $s > p$. Figure **f** shows a double-logarithmic plot of the global error versus the cpu time [92, 94]

6.5.4 The Woodpecker Toy

The woodpecker toy is a well known benchmark example of a non-smooth system, see for example [39, 44]. The toy consists of a pole, a sleeve with a hole that is slightly larger than the diameter of the pole, a spring and the woodpecker body. In operation, the woodpecker moves down the pole performing some kind of pitching motion, which is controlled by the sleeve [44]. The woodpecker of Glocker [39, 44] is modeled by six set-valued laws: A unilateral contact and friction element between beak and pole, a unilateral contact and friction element between the lower end of the sleeve and the pole and a unilateral contact and a friction element between the upper end of the sleeve and the pole. The mechanical model as well as the initial conditions for the simulation can be found in [44]. The following simulation parameters have been chosen: Minimal step size $\Delta t_{min} = 10^{-7}$s, maximal step size $\Delta t_{max} = 5 \cdot 10^{-3}$s, maximal integration order $p_{max} = 6$. Figure 6.10a depicts the model of the woodpecker [39]. Figures 6.10b shows the the phase diagram φ_S versus $\dot{\varphi}_S$. The time evolution of $\dot{\varphi}_M$, $\dot{\varphi}_S$, Δt and of the used integration order are depicted in Figs. 6.10c-f. Around the time instance $t = 0.1s$ high oscillations are present and the step size does not reach Δt_{max}, because otherwise the integration order would be larger than the maximal user defined order. The time evolution of $\dot{\varphi}_M$ and $\dot{\varphi}_S$ are complemented by a diagram which shows the projection modus of the six different discrete set-valued laws. Black parts indicate "constraint" character, gray parts "impressed" character, respectively. Table 6.1 lists the different switching incidents. In analogy to the example of the single DOF impact oscillator in Sect. 6.5.3, the time evolution of the woodpecker has also been computed with a step size adjustment module which uses a fixed integration order. In contrast to the single DOF impact oscillator, the same minimal step size $\Delta t_{min} = 10^{-10}$s has been chosen for all computations. The different approximations have been compared with a reference solution which has been computed using an absolute error tolerance of $atol_{\mathrm{u}} = 10^{-10}$ m/s, a minimal step size $\Delta t_{min} = 10^{-10}$s and a maximal step size $\Delta t_{max} = 9 \cdot 10^{-4}$s. In Fig. 6.11a the double-logarithmic plot of the maximal step size Δt_{max} versus the

Table 6.1 The different switching incidents of the woodpecker toy

Set-valued law		from	to
Friction element	2	slip	stick
Friction element	2	stick	slip
Unilateral contact	2	closed	open
Unilateral contact	3	open	closed
Unilateral contact	3	closed	open
Unilateral contact	1	open	closed
Unilateral contact	1	closed	open
Unilateral contact	3	open	closed
Unilateral contact	3	closed	open
Unilateral contact	2	open	closed
Unilateral contact	2	closed	open
Unilateral contact	2	open	closed

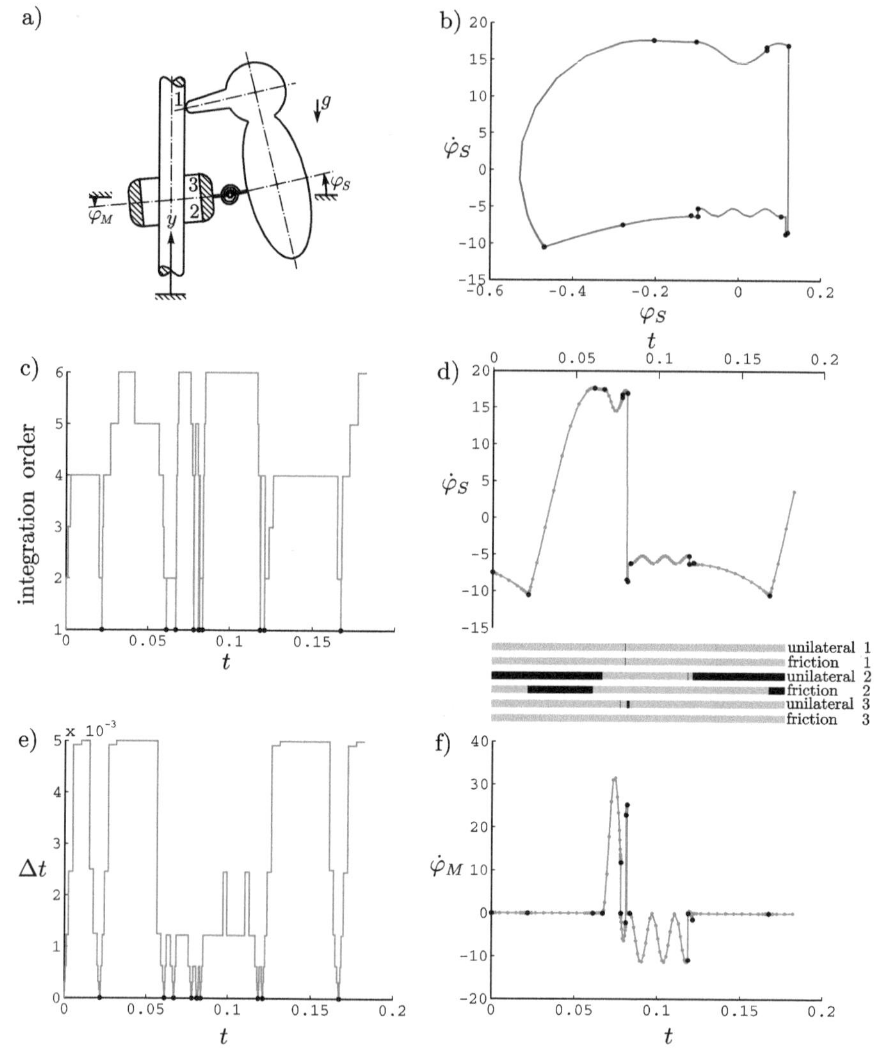

Fig. 6.10 Woodpecker toy together with the results of the simulation. The time evolution of $\dot{\varphi}_M$ and $\dot{\varphi}_S$ are complemented by a diagram which shows the projection modus of the six different discrete set-valued laws. *Black* parts correspond to "constraint" character, *gray* parts to "impressed" character, respectively [92, 94]

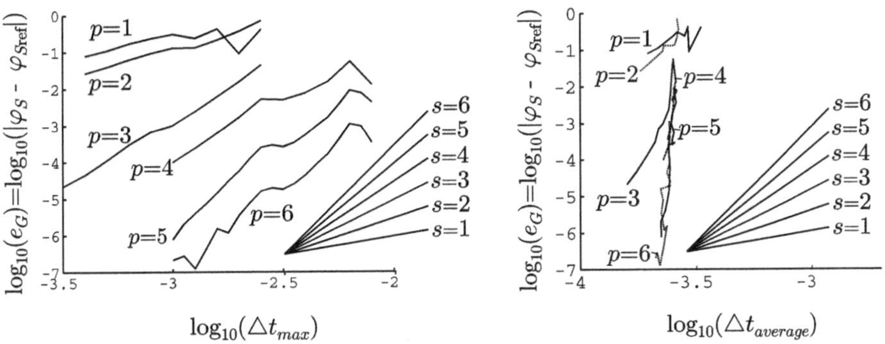

Fig. 6.11 Double-logarithmic plot of the maximal step size Δt_{max} and the average step size $\Delta t_{average}$ versus the the global integration error e_G of the coordinate φ_S [94]

global integration error e_G of the coordinate φ_S is depicted. Note that the double-logarithmic plot consists of more or less straight lines with slope $s = p$, thus even the use of a constant minimal step size Δt_{min} yields an integration order of p with respect to Δt_{max}. A possible explanation might be that the few local errors of the switching intervals can be neglected compared to the accumulation of the many local errors during the smooth parts of the motion. In addition, the approximations are compared with a reference solution which also uses the same minimal step size $\Delta t_{min} = 10^{-10}$s. Figure 6.11b depicts the double-logarithmic plot of the average step size $\Delta t_{average}$ versus the global integration error e_G.

Chapter 7
The dynamY Software

This chapter provides a short overview on the dynamY software package [88], which can be used to simulate non-smooth mechanical systems. The dynamY software package consists of various C++ classes, which allow for the modeling and the simulation of a non-smooth system. The software works on the basis of smooth mechanical systems, which can be connected by non-smooth elements. The chapter gives a short overview on the software and discusses some calculated examples[88].

7.1 Overview

The dynamY environment defines objects which model the mechanical system and objects which act on this system, i.e. which perform the simulation. All these different objects are managed by administration classes. The description of the mechanical setup is lead by the following philosophy: The entire mechanical system consists of mechanical subsystems,

$$\begin{aligned} \mathbf{M}_i \mathrm{d}\mathbf{u}_i - \mathbf{h}_i \mathrm{d}t &= 0, \\ \mathrm{d}\mathbf{q}_i - \mathbf{F}_i \mathbf{u}_i \, \mathrm{d}t &= 0, \end{aligned} \tag{7.1}$$

which are connected by non-smooth elements. These mechanical subsystems model for example the movement of a rigid body in space but also more specific systems, for example a pendulum in minimal coordinate form. A non-smooth element incorporates two aspects: On the one hand the non-smooth element i defines the generalized force direction $\mathbf{W}_i$ etc., i.e. geometrical (and kinematical) aspects. On the other hand, a non-smooth element i incorporates a set-valued law i. Active set-valued laws are considered on velocity level by the inclusion

$$-(\gamma_{Ei} + \varepsilon_i \gamma_{Bi}) \in \mathscr{N}_{\hat{\mathscr{A}}_i}(\hat{\mathbf{P}}_i), \tag{7.2}$$

which is already in discrete form. Non-active geometric discrete set-valued laws, i.e. geometrical discrete set-valued laws with $\mathbf{g}_i \neq 0$ in a discrete sense, can not be described by (7.2). Such laws are in impressed mode and are considered in the

C. Studer: Numerics of Unilateral Contacts and Friction, LNACM 47, pp. 129–159.
springerlink.com

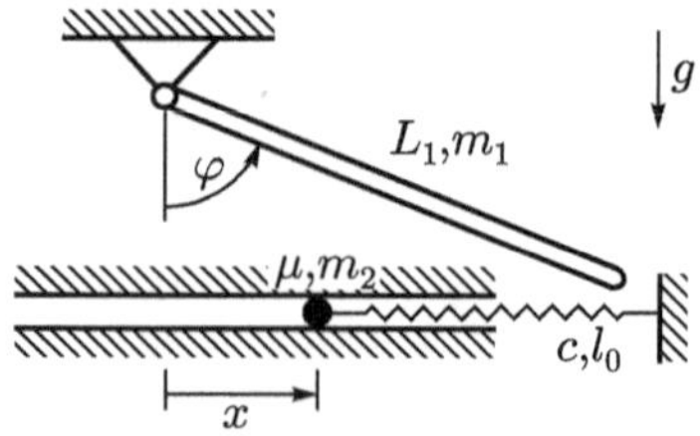

Fig. 7.1 Example of a non-smooth system, which consists of a planar physical pendulum and a point mass

same way as the vector of gyroscopic and external forces $\mathbf{h}$. Furthermore, impressed forces like springs, dampers or other external forces may act on or between the mechanical subsystems.

The different mechanical subsystems are represented by objects which are derived from an abstract base class pms (partial mechanical system). In the same manner, all the different non-smooth elements are derived from an abstract base class nsc (non-smooth constraint) and all the different external forces are derived from an abstract base class gef (generalized external forces). The derived pms objects assemble the mass matrix $\mathbf{M}$, the external and gyroscopic terms $\mathbf{h}$ and the matrix $\mathbf{F}$, which links $\dot{\mathbf{q}}$ to $\mathbf{u}$. The derived nsc objects assemble the generalized force directions $\mathbf{W}$, the inhomogeneity term ζ and define the associated discrete set-valued laws. Furthermore, the nsc objects determine whether the modeling on velocity level by (7.2) is applicable or not, i.e. the objects evaluate whether the associated discrete set-valued law is active or not. The derived gef objects add external forces directly to the vector of external and gyroscopic forces $\mathbf{h}$. Note in this context that external forces may either be assembled by the pms objects or by the gef objects. The latter option must be chosen if external forces are applied between different pms objects, take for instance a spring which acts between two bodies.

Consider a mechanical system, which consists of a planar physical pendulum and a point mass, see Fig. 7.1. The displacement of the point mass, which is restricted to the horizontal x-axis by a bilateral constraint, is subjected to friction. In addition, a unilateral contact is assumed between the pendulum and the point mass. Furthermore, a spring acts between the point mass and the inertial surrounding. The pendulum and the point mass are modeled by the derived pms-classes pmsPend2D and pmsPom2D. The unilateral and the frictional bilateral constraint are represented by the derived nsc objects nscPend2DPom2DUni and nscPom2DInertBilFric, respectively. The spring is characterized by the derived gef object gefPom2DInertSpring. The derived classes which characterize the non-smooth system are shown together with their abstract base classes in Fig. 7.2. Furthermore, Table 7.1 depicts how the different derived pms, nsc and gef objects assemble the necessary equations.

The different derived pms, nsc and gef objects are managed by the administrative mecS-class. This class includes three lists with base class pointers to all derived pms, nsc and gef objects. The mecS-class of the example can be seen in Fig. 7.4.

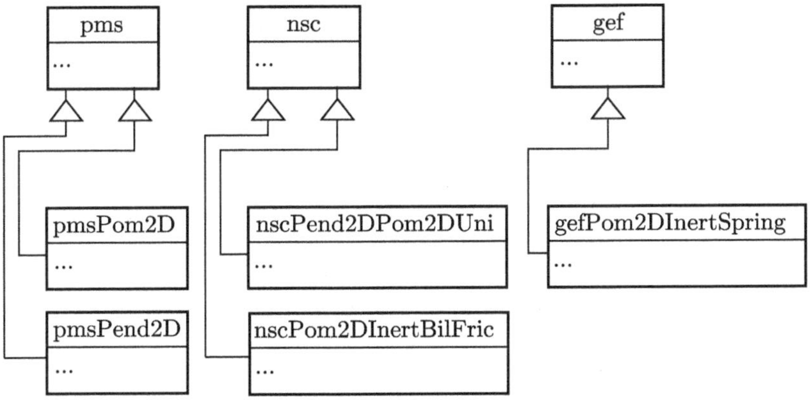

Fig. 7.2 Abstract pms, nsc and gef base classes together with their child objects. The classes pmsPend2D and pmsPom2D inherit from the abstract base class pms, the classes nscPend2DPom2DUni and nscPom2DInertBilFric inherit from the abstract base class nsc, and the class gefPom2DInertSpring inherits form the abstract base class gef

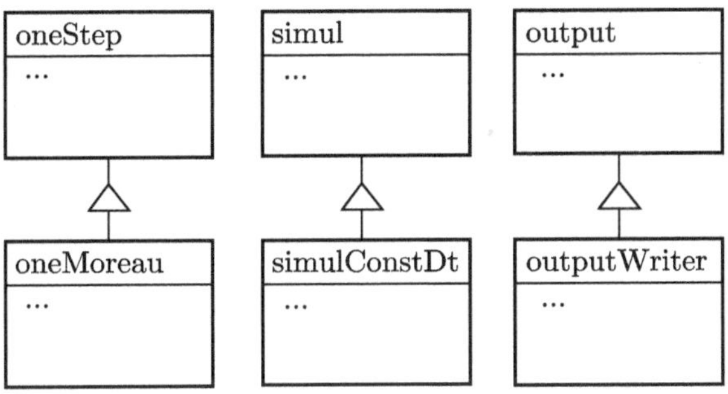

Fig. 7.3 The simul, oneStep and output abstract base classes with derived objects

So far, the modeling of a mechanical system has been discussed. The aim of the dynamY object structure is to do something with this mechanical system, more precisely to simulate the mechanical behaviour of the system in time. All objects derived from an abstract base class oneStep are intended to perform one time step based on the mechanical system. The dynamY objects are mainly designed to apply a Moreau time step on the mechanical system, where the discrete set-valued laws are solved iteratively. Nevertheless, other solution methods can of course also access the mechanical system, but the implementation might become more complicated, and some parts of the dynamY object structure might become useless.

In order to do a whole simulation, the time steps which have been performed by the derived oneStep object are stringed together. This is done by an object which is derived from the abstract base class named simul. Furthermore, a child object which is derived from an abstract base class output handles the data output. Take for instance a Moreau based one step solver oneMoreau derived from the abstract base class oneStep, a simulation class simulConstantDt derived from the abstract base class simul and a derived output object outputWriter, see also Fig. 7.3.

The root class is the main administration class, which contains a pointer to the mechanical system mecS as well as base class pointers to the derived oneStep, simul and output object. All these three derived objects inherit from their mother a pointer back to the root. Also the mecS-class has a pointer back to the root object. The communication between the objects takes always place via the root object: It is supposed that the derived simul object communicates with the derived output and the derived oneStep object, and that only the derived oneStep object accesses the mechanical system mecS. Figure 7.4 depicts the root object together with the mecS object, the derived oneStep object oneMoreau, the derived simul object simulConstantDt and the derived output object outputWriter.

The idea of the dynamY object structure is to provide an object organization which enables the user to implement and solve his mechanical problem in a proper way. The whole administration, the abstract base classes and some derived oneStep, simul and output objects are provided by dynamY . All these classes are gathered in the dynamYKernelLib library. It is the user's task to derive own pms, nsc and gef child classes in order to describe his problem. This gives the user maximal flexibility in modeling his system. We tried to make the structure of the pms, nsc and gef abstract base classes as simple as possible to make this part of dynamY also accessible to non-C++ cracks. We also tried to keep all information together in these three types of classes to avoid a class jungle. The dynamYAppLib library is intended to gather all derived pms, nsc and gef classes. It is of course also possible to derive own oneStep, simul and output classes, but these are of course more complicated to implement.

pms		nsc		gef	
pmsPom2D	pmsPend2D	nscPend2DPom2DUni	nscPom2DInertBilFric	gefPom2DInertSpring	
$\mathbf{M}_1 \mathrm{d}\mathbf{u}_1 - \mathbf{h}_1 \mathrm{d}t$		$-\mathbf{W}_{11} \mathrm{d}\hat{\mathbf{P}}_1$	$-\mathbf{W}_{12} \mathrm{d}\hat{\mathbf{P}}_2$	$-\mathbf{f}_{11} \mathrm{d}t$	$= 0$
	$\mathbf{M}_2 \mathrm{d}\mathbf{u}_2 - \mathbf{h}_2 \mathrm{d}t$	$-\mathbf{W}_{21} \mathrm{d}\hat{\mathbf{P}}_1$			$= 0$
$\dot{\mathbf{q}}_1 - \mathbf{F}_1 \mathbf{u}_1$					$= 0$
	$\dot{\mathbf{q}}_2 - \mathbf{F}_2 \mathbf{u}_2$				$= 0$
		$-(\gamma_{E1} + \varepsilon_1 \gamma_{B1}) \in \mathscr{N}_{\hat{\mathscr{A}}_1}(\hat{\mathbf{P}}_1)$			
			$-(\gamma_{E2} + \varepsilon_2 \gamma_{B2}) \in \mathscr{N}_{\hat{\mathscr{A}}_2}(\hat{\mathbf{P}}_2)$		

Table 7.1 Necessary equations to describe the example depicted in Fig. 7.1. The derived pms, nsc and gef-classes add their mechanical subsystem, their non-smooth element and their generalized external force to the mechanical system

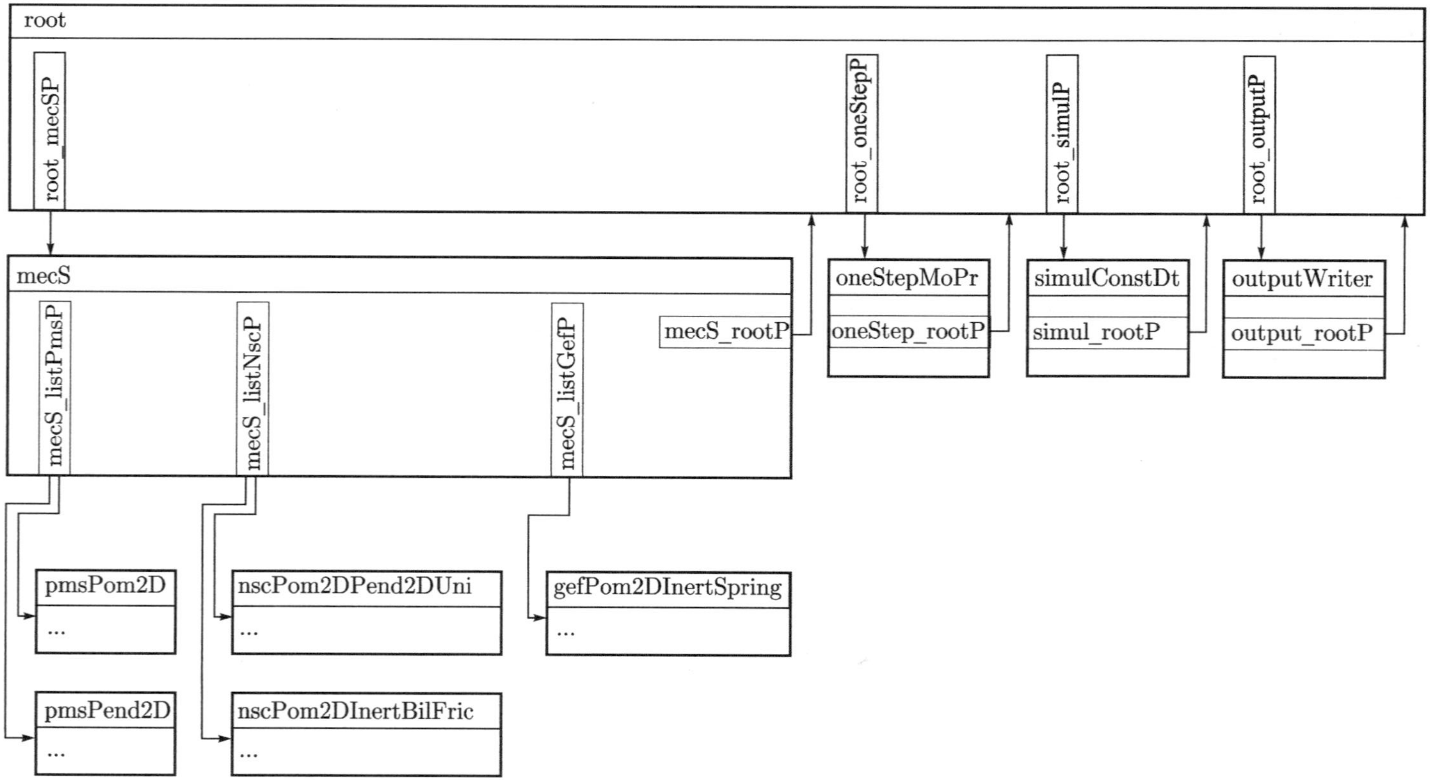

Fig. 7.4 The root object is the main administrative object and holds a pointer to the mecS object as well as base class pointers to the derived oneStep, simul and output object. The mecS objects itself holds base class pointers to the derived pms, nsc and gef classes

7.2 Modeling a Partial Mechanical System

A partial mechanical system consists of a mass matrix $\mathbf{M}_i$, a vector of gyroscopic and external forces $\mathbf{h}_i$ and of a matrix $\mathbf{F}_i$, according to equation (7.1).

The pms abstract base class provides the member variables pms_numQ, pms_numU, pms_posQ and pms_posU. These variables specify the number of degrees of freedom with respect to the positions and the velocities, and define the position of the object's first degree of freedom within the global vectors $\mathbf{q}$ and $\mathbf{u}$. In addition, the pms base class provides the interface to the pure virtual methods listed in Table 7.2. The implementation of these functions must be done in the derived classes.

Table 7.2 Pure virtual methods of the pms abstract base class

method name	description
pms_assembleH($\mathbf{h}$,$\mathbf{q}$,$\mathbf{u}$,t)	assembles the object's $\mathbf{h}_i$ into $\mathbf{h}$.
pms_assembleInvM($\mathbf{M}^{-1}$,$\mathbf{q}$,$\mathbf{u}$,t)	assembles the object's $\mathbf{M}_i^{-1}$ into $\mathbf{M}^{-1}$
pms_assembleF($\mathbf{F}$,$\mathbf{q}$,$\mathbf{u}$,t)	assembles the object's $\mathbf{F}_i$ into $\mathbf{F}$
pms_geom($\mathbf{q}$)	performs some geometrical precalculations which can be used by the assembly methods pms_assembleH, pms_assembleInvM and pms_assembleF

Note that the pure virtual method pms_assembleInvM assembles the inverse mass matrix $\mathbf{M}^{-1}$. As the global matrix $\mathbf{M}$ has bloc structure, each of these blocs can be inverted separately. The method pms_assembleInvM first builds the inverse of $\mathbf{M}_i$ analytically or numerically and assembles it afterwards to $\mathbf{M}^{-1}$.

For some mechanical systems, time consuming operations on the displacements $\mathbf{q}_i$ have to be performed, for example when calculating rotation matrixes. If several assemble methods request the same geometrical operations, then it is useful to do these operations only once, within the pms_geom method. This method is called by the oneStep solver prior to the assemble methods pms_assembleH, pms_assembleInvM and pms_assembleF, and the same vector $\mathbf{q}_i$ is passed to the pms_geom and the assemble methods in order to avoid a mess.

Furthermore, there are two virtual methods pms_reviewQ and pms_reviewU which take no action if not re-implemented in the derived pms-classes. These methods are intended to alter the global vectors $\mathbf{q}$ and $\mathbf{u}$. This might be useful if some projections are needed, for example if Euler parameters are used for angular parametrization. It is also possible to use these two methods for kinematical excitation of degrees of freedoms. In this case, the dynamics must be switched off by choosing the associated masses or inertias infinitely large.

7.3 Modeling a Non-Smooth Element

As already mentioned in Sect. 7.1, a non-smooth element defines methods which are associated with geometric aspects (calculation of the generalized force direction $\mathbf{W}_i$, inhomogeneity term ζ_i, etc.), or with the discrete set-valued law (7.2). The latter methods are strongly related to the algorithm which is used to solve the inclusion problem. If an augmented Lagrangian approach is used, then the methods have to offer a projection to the discrete convex set $\hat{\mathscr{A}}_i = \mathscr{A}_i \Delta t$. If the inclusions are solved by a linear complementarity approach, then the methods must perform the appropriate decomposition. It would indeed be possible to separate the methods associated with the discrete set-valued laws form the geometrical methods by introducing a discrete set-valued law object, which is connected to a pure geometric nsc object by a pointer. However, this separation generates administrative effort, data exchange and downcasting operations on pointers. In order to keep all information together, the program system dynamY incorporates both the geometric and the discrete set-valued law methods in one nsc object. The methods associated with the discrete set-valued law are designed to assist an iterative augmented Lagrangian approach. The software dynamY is intended to express the discrete set-valued laws on velocity level by (7.2), which offers the following advantages:

- all discrete set-valued laws can be treated in the same way by the solver (oneStep object),
- the corresponding differential algebraic system has index two instead of three,
- partially elastic impact and non-impulsive motion can be expressed with the same discrete set-valued law,
- the gap function $\mathbf{g}(\mathbf{q},t)$ must not be evaluated during the iterative solution process,
- non-active geometric discrete set-valued laws have not to be considered during the iterative solution process.

The main disadvantage is the occurring drift problem.

A geometric discrete set-valued law can only be treated on velocity level by (7.2) if $\mathbf{g} = 0$ in a discrete sense, i.e. if the geometric discrete set-valued law is active. The dynamY software makes the following distinction:

- A level flag 2 is assigned to non-active geometric discrete set-valued laws for which $\mathbf{g}_i \neq 0$ and $\hat{\mathbf{P}}_i = 0$. It is assumed that the associated discrete percussions $\hat{\mathbf{P}}_i$ vanish, i.e. non-smooth elements with level flags 2 are neglected by the one-step solver. An example is an open unilateral contact i with discrete percussions $\hat{P}_i$ equal to zero.
- A level flag 1 is assigned to non-active geometric discrete set-valued laws for which $\mathbf{g}_i \neq 0$ but $\hat{\mathbf{P}}_i \neq 0$. Such discrete set-valued laws are treated in the same way as the vector $\mathbf{h}$, for example a pre-stressed spring. A non-smooth element with level flag 1 must provide information about the impressed discrete percussions $\hat{\mathbf{P}}_i$ and the generalized force direction $\mathbf{W}_i$. Furthermore, the value of the gap function $\mathbf{g}(\mathbf{q},t)$ is needed.

- A level flag 0 is assigned to all active discrete set-valued laws which can be modeled on velocity level by (7.2), i.e. all bilateral constraints, all kinematic discrete set-valued laws and all active geometric discrete set-valued laws. Take for instance a closed unilateral contact, a friction element or a joint. The treatment of a non-smooth element with level flag 0 requests information about the generalized force direction $\mathbf{W}_i$, the inhomogeneity term ζ_i and the matrix ε_i. Furthermore, a projection function to the discrete convex set $\hat{\mathscr{A}_i}$ which characterizes the discrete set-valued law (7.2) is needed.

To conclude, the nsc object should offer a method to determine the level flag of the discrete set-valued law. Note in this context that the criterion $\mathbf{g}_i = 0$ has to implemented in a discrete sense. Take for instance the unilateral contact: a possible criterion would be $g_i \leq 0$. The possible criterion for a pre-stressed spring is $|g_i| \leq tol$ with tol being a user defined tolerance. The nsc object should also define methods which assemble the generalized force directions $\mathbf{W}$, the homogeneity term ζ and the matrix ε. In addition, there are two methods which are associated with the discrete set-valued law: a method for non-smooth elements with level flag 1 which provides the discrete percussions $\hat{\mathbf{P}}_i$ in case of $\mathbf{g}_i \neq 0$, and a method for non-smooth elements with level flag 0, which defines the projection to the discrete convex set $\hat{\mathscr{A}_i}$.

In the following, the nsc abstract base class is discussed. The member variables nsc_numLa, nsc_posLa, and nsc_internalPosLa define the dimension of the discrete percussion $\hat{\mathbf{P}}_i$ and its position in the global vector $\hat{\mathbf{P}}$. While the variable nsc_internalPosLa refers to the position within the vector $\hat{\mathbf{P}}$ which incorporates all non-smooth elements, the variables nsc_posLa indicates the position in a vector $\hat{\mathbf{P}}$ of selected non-smooth elements, i.e. non-smooth elements with level flag 1 or 0. This latter position should be used by all assembly methods. A further member variable is nsc_subProblems, which is only applicable if several discrete set-valued laws are gathered in one nsc object. The variable defines the number and the dimension of the subproblems, which allows the solver to optimize the choice of $\mathbf{R}_i$.

The pure virtual methods of the nsc abstract base class are listed in Table 7.3. When dealing with complex geometric systems, time-consuming geometric calculations have to be made to compute the generalized force directions $\mathbf{W}_i$, the gap function $\mathbf{g}_i$ and the level flag of the non-smooth element. In order to avoid repeating all these calculations in the methods nsc_assembleW, nsc_assembleZeta, nsc_assembleEps, nsc_assembleG and nsc_determineLevel, they can be done in the method nsc_geom. Note that these precalculations require that the same global vector $\mathbf{q}$ is passed to nsc_geom, nsc_assembleW, nsc_assembleZeta, nsc_assembleEps, nsc_assembleG and nsc_determineLevel, and that nsc_geom is called prior to the other methods using the results of nsc_geom.

The nsc abstract base class provides additional virtual methods which take no action if not re-implemented in the derived classes. In analogy to the pms review methods, the nsc abstract base class offers a virtual method nsc_reviewLa, which might be useful if the non-smooth element wants to make an initial guess for its discrete percussion $\hat{\mathbf{P}}_i$. A further method is the nsc_assembleDriftSt method. This method allows to add an additional drift stabilization term to (7.2), i.e.

Table 7.3 Pure virtual methods of the nsc abstract base class

method name	description
nsc_assembleW($\mathbf{W}$,$\mathbf{q}$,$\mathbf{u}$,t)	assembles the object's $\mathbf{W}_i$ to the global $\mathbf{W}$
nsc_assembleZeta(ζ,$\mathbf{q}$,$\mathbf{u}$,t)	assembles the object's ζ_i to the global ζ
nsc_assembleEps (ε,$\mathbf{q}$,$\mathbf{u}$,t)	assembles the object's ε_i to the global ε
nsc_assembleG($\mathbf{g}$,$\mathbf{q}$,$\mathbf{u}$,t)	assemble the object's $\mathbf{g}_i$ to the global $\mathbf{g}$
nsc_determineLevel($\mathbf{q}$,$\mathbf{u}$,t)	determines level flag
nsc_geom($\mathbf{q}$)	performs time-consuming geometric pre-calculations for the methods nsc_assembleW, nsc_assembleZeta, nsc_assembleEps, nsc_assembleG and nsc_determineLevel.
nsc_lawV($\mathbf{x}_{in}, \mathbf{x}_{out}, \sigma$)	performs the projection to the convex set $\hat{\mathscr{A}}_i$
nsc_initLawV($\mathbf{q}$,$\mathbf{u}$,t,γ,Δt)	initializes the method nsc_lawV
nsc_lawD($\mathbf{x}_{out}, \sigma$)	returns the impressed $\hat{\mathbf{P}}_i$ for $\mathbf{g} \neq 0$
nsc_initLawD($\mathbf{q}$,$\mathbf{u}$,t,$\mathbf{g}$,Δt)	initializes the method nsc_lawD

$$-\left(\frac{g_D}{\Delta t} + \gamma_{Ei} + \varepsilon\gamma_{Bi}\right) \in \mathscr{N}_{\hat{\mathscr{A}}_i}(\hat{\mathbf{P}}_i). \tag{7.3}$$

It is the solver's task to determine at which time instance D the stabilization term is evaluated, as this time instance must be adjusted to the position update formula used. Using Moreau's midpoint rule, the stabilization term must be evaluated at $\mathbf{q}_M$. In either case, the drift stabilization term is evaluated in an explicit way. Note that drift stabilization requires $\varepsilon = 0$. It is mainly advised to use this stabilization only for bilateral constraints which are affected from heavy drift. Note that the drift stabilization term $\frac{g_M}{\Delta t}$ can easily incorporated into Moreau's time-stepping scheme by just adding it to the constant vector $\mathbf{c}$ in (5.53), i.e.

$$\mathbf{c}_i = (\mathbf{I} + \varepsilon)\gamma_{Bi} + \mathbf{W}_{Mi}^{\top}\mathbf{M}_M^{-1}\mathbf{h}_M\Delta t + \zeta_{Mi} + \frac{g_M}{\Delta t}, \quad \varepsilon = 0. \tag{7.4}$$

7.4 Modeling External Forces

All external forces including springs and dashpots are directly considered in the $\mathbf{h}$-vector. No damping and stiffness matrices are specified. External forces which act only on one partial mechanical system can be added to the $\mathbf{h}$-vector by the pms_assembleH method. External forces which act between partial mechanical systems like springs and dashpots can be considered by derived objects of a abstract base class gef (generalized external force). These derived gef objects hold pointers to the corresponding pms objects, can access their information and alter the corresponding parts in the $\mathbf{h}$-vector. For this purpose the gef abstract base class defines a pure virtual method gef_assembleH.

Table 7.4 Available derived oneStep objects

class	incl-solver	abort	$\mathbf{u}$
oneMoPrJORAbortLaExp	JORprox	in $\hat{\mathbf{P}}$	explicit
oneMoPrSORAbortLaExp	SORprox	in $\hat{\mathbf{P}}$	explicit
oneMoPrSORAbortUExp	SORprox	in $\mathbf{u}$	explicit
oneMoPrSORAbortUImpU	SORprox	in $\mathbf{u}$	implicit

Table 7.5 Available derived simul-classes

class	description
simulConstantDt	performs simulation with constant step size Δt
simulAdjustDtExV2	performs simulation with step size adjustment and higher integration order in smooth time steps. Extrapolation uses only uneven number of substeps. Error estimation steers the integration order
simulAdjustDtExV1	performs simulation with step size adjustment and higher integration order in smooth time steps. Extrapolation uses even and uneven number of substeps. Error estimation steers the integration order
simulAdjustDtExFixOrderV2	performs simulation with step size adjustment and higher integration order in smooth time steps. Extrapolation uses only uneven number of substeps. The integration order of smooth time steps is defined by the user

7.5 One Step and Simulation Objects

Objects derived from the abstract base class oneStep are expected to perform one time step with the mechanical system mecS. The available derived oneStep objects are listed in Table 7.4. The column "incl-solver" lists the kind of iterators which are used to solve the inclusions. The column "abort" states in which variables the abort criterion of the iteration is evaluated. The column $\mathbf{u}$ indicates whether the velocities in the vector of gyroscopic and external forces $\mathbf{h}$ and in the inhomogeneity term ζ are considered explicitly or implicitly. All objects use Moreau's midpoint method as time-stepping scheme. Furthermore, all calculations are explicit with respect to the positions $\mathbf{q}$.

An object which is derived from the abstract base class simul strings together the different time steps. The available derived simul-classes are listed in Table 7.5.

7.6 Examples

In this section some examples are presented which have been calculated with the dynamY software package. The aim of this section is to give the reader an idea how different mechanical problems can be approached by using the dynamY package.

Additional examples which focus on the topic "augmented time-stepping by step size adjustment and extrapolation" can be found in Sect. 6.5.

7.6.1 Bodies in 3D Space

Rigid Bodies in the three-dimensional space can be parameterized by the positions $\mathbf{q} = (x\, y\, z\, e_0\, e_1\, e_2\, e_3)^\top$ and the velocities $\mathbf{u} = (u_x\, u_y\, u_z\, \omega_x\, \omega_y\, \omega_z)^\top$, where $\mathbf{e} = (e_0\, e_1\, e_2\, e_3)^\top$ are the Euler parameters for angular parametrization, i.e.

$$e_0 = \cos\frac{\chi}{2}, \quad e_1 = n_1 \sin\frac{\chi}{2}, \quad e_2 = n_2 \sin\frac{\chi}{2}, \quad e_3 = n_3 \sin\frac{\chi}{2}. \tag{7.5}$$

The vector ${}_I\mathbf{n} = (n_1\, n_2\, n_3)^\top$ is the (unit) rotation axis evaluated in the inertial system I. The angle χ parameterizes the rotation around this rotation axis. Note that $||\mathbf{e}||_2 = 1$, i.e. if Euler parameters are used within a discrete scheme, then a projection $\mathbf{e} \to \frac{\mathbf{e}}{||\mathbf{e}||_2}$ must be applied from time to time to prevent drift. Considering the dynamY package, this projection is implemented in the virtual pms method pms_reviewQ, which is called by the oneStep solver after each update of $\mathbf{q}$.

The components ${}_I\mathbf{r}_S = (x y z)^\top$ of $\mathbf{q}$ and ${}_I\mathbf{u}_S = (u_x\, u_y\, u_z)^\top$ of $\mathbf{u}$ denote the position and the velocity of the body's center of mass S in the inertial system. The rotational velocity components ${}_{I/B}\boldsymbol{\Omega} = (\omega_x\, \omega_y\, \omega_z)^\top$ of $\mathbf{u}$ may either be evaluated in the inertial or body-fixed coordinate system I or B, where the latter is advisable if the inertia tensor Θ is not of type sphere, i.e. if ${}_I\Theta$ is not constant. The equations of motion are obtained by

$$m\mathbf{I}_I(\dot{\mathbf{u}}_S) = {}_I\mathbf{F}, \tag{7.6}$$

$${}_{I/B}\Theta_{S\,I/B}(\dot{\Omega}) + {}_{I/B}\Omega \times ({}_{I/B}\Theta_{S\,I/B}\Omega) = {}_{I/B}\mathbf{M}_S, \tag{7.7}$$

with $\mathbf{F}$ and $\mathbf{M}$ being the external force and the external moment acting on the body's center of mass S. The relation between $\dot{\mathbf{q}}$ and $\mathbf{u}$ is given by $\mathbf{F}_i \in \mathbb{R}^{7\times 6}$

$$\dot{\mathbf{q}}_i = \mathbf{F}_i\mathbf{u} \quad \begin{cases} \mathbf{F}_i = \begin{pmatrix} \mathbf{I} & 0 \\ 0 & \frac{1}{2}\mathbf{H}^\top \end{pmatrix} & \text{for } {}_I\Omega \text{ in inertial system } I, \\ \mathbf{F}_i = \begin{pmatrix} \mathbf{I} & 0 \\ 0 & \frac{1}{2}\bar{\mathbf{H}}^\top \end{pmatrix} & \text{for } {}_B\Omega \text{ in body-fixed system } B, \end{cases} \tag{7.8}$$

with the matrices $\mathbf{H}$ and $\bar{\mathbf{H}}$ defined by

$$\mathbf{H} = \begin{pmatrix} -e_1 & e_0 & -e_3 & e_2 \\ -e_2 & e_3 & e_0 & -e_1 \\ -e_3 & -e_2 & e_1 & e_0 \end{pmatrix}, \qquad \bar{\mathbf{H}} = \begin{pmatrix} -e_1 & e_0 & e_3 & -e_2 \\ -e_2 & -e_3 & e_0 & e_1 \\ -e_3 & e_2 & -e_1 & e_0 \end{pmatrix}. \tag{7.9}$$

The transformation matrix ${}_I\mathbf{x} = \mathbf{A}_{IB\,B}\mathbf{x}$, which might be useful for the calculation of the generalized force directions $\mathbf{W}_i$, is

$$\mathbf{A}_{IB} = \mathbf{H}\bar{\mathbf{H}}^\top = \begin{pmatrix} 2(e_0^2+e_1^2)-1 & 2(e_1e_2-e_0e_3) & 2(e_1e_3+e_0e_2) \\ 2(e_1e_2-e_0e_3) & 2(e_0^2+e_2^2)-1 & 2(e_2e_3-e_0e_1) \\ 2(e_1e_3-e_0e_2) & 2(e_2e_3+e_0e_1) & 2(e_0^2+e_3^2)-1 \end{pmatrix}. \tag{7.10}$$

For more detailed information on the topic discussed above the reader is referred to [54]. In the following, some examples of rigid bodies in three-dimensional space are discussed.

7.6.1.1 One Ball on a Plane

Consider a ball on a plane, which is spanned by the x- and y-axis of the inertial system I, see Fig. 7.5a. All vectors are evaluated in this inertial system I. The ball has an initial velocity $u_y = 1$ m/s in the y-direction and and initial spin $\omega_z = 50$ rad/s around the z-axis. Between ball and plane acts a unilateral contact, where additionally Coulomb-Contensou and rolling friction is present. The following parameters have been used for the simulation: $r = 2.5$ cm, $m = 510.5$ g, and $\Theta_{x,y,z} = 1.276 \cdot 10^{-4}$ kg m^2. The discrete percussion $\hat{P}_N$ is associated with the unilateral contact. Considering the Coulomb-Contensou friction, a maximal Coulomb friction force of $0.2 \cdot \hat{P}_N$ and a maximal Contensou drilling torque $\tau_{ccz,max} = 0.00665\text{m} \cdot \hat{P}_N$ has been chosen. Note that the maximal drilling torque is chosen large to underline its action. In order to make the values of $\mathbf{G}_{ii}$ similar, it is advised to scale the drilling torque by the radius r of the sphere, i.e. to work with the force $\lambda_{ccz} = \tau_{ccz}/r$. The rolling friction torque is described by a normal cone inclusion with respect to the set $\mathscr{S}_r = \{\xi \in \mathbb{R}^2 \mid ||\xi||_2 \leq 0.002\text{m}\hat{P}_N\}$, see for example [78]. Figures 7.5a-f depict the results of the simulation. Two switching points occur, the first at $t = 0.31$s at which a slip-stick transition occurs, the second at $t = 1.27$s at which the ball comes to a rest. Figure 7.5b depicts the rolling friction torque τ_{rx} around the x-axis. The torque vanishes at the second switching incident at which the ball comes to a rest. The Coulomb-Contensou friction forces $\lambda_{ccz} = \tau_{ccz}/r$ and λ_{ccy} which are associated with the rotation around the z-axis and with the translation in y direction are depicted in Figs. 7.5c and d. The time evolutions of the associated rotations ω_z and ω_x can be seen in Figs. 7.5e and f. The angular velocity ω_z decreases due to λ_{ccz}. As a consequence, the relative velocity in y-direction becomes dominant, which causes $|\lambda_{ccz}|$ to decrease and $|\lambda_{ccy}|$ to increase according to the Coulomb-Contensou law. The angular velocity $|\omega_x|$ increases. At the first switching incident at $t = 0.31$s the Coulomb-Contensou law switches to sticking mode. The rolling friction torque τ_{rx} is still in impressed mode, which causes the angular velocity $|\omega_x|$ to decrease again until the ball comes to a rest.

7.6.1.2 Two Balls on a Plane

Consider now two balls on a plane, see Fig. 7.6a. Between the two balls acts a frictional unilateral contact with normal restitution and friction coefficient $\varepsilon_N = 1$ and

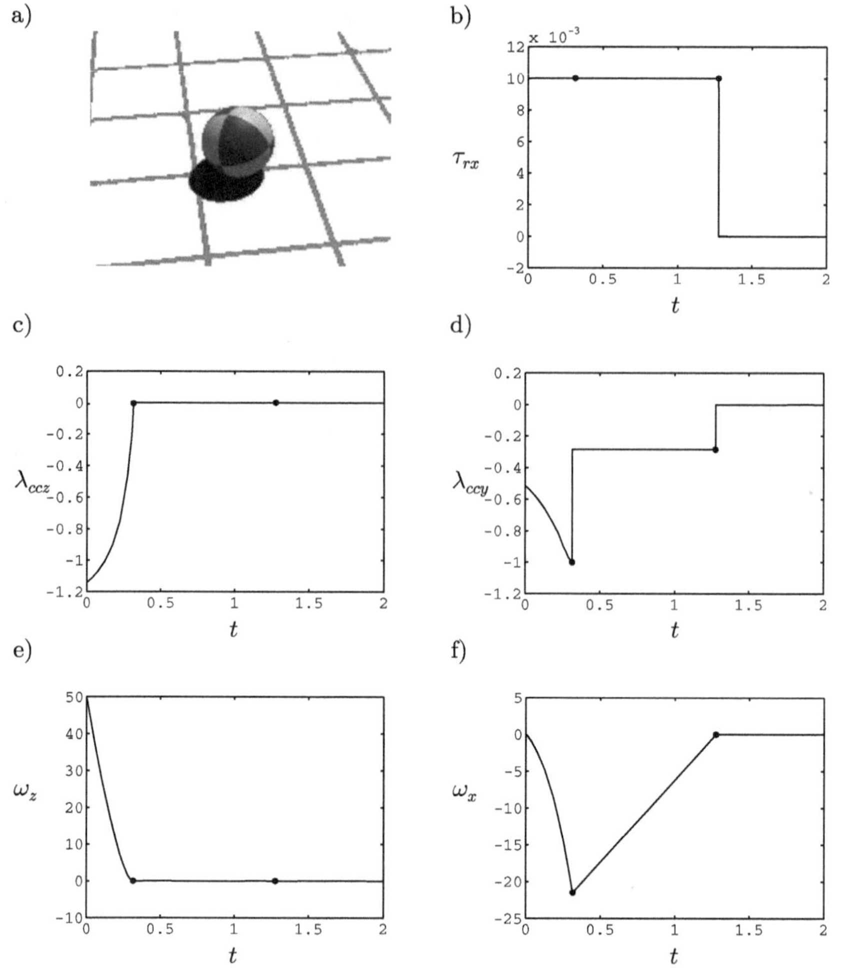

Fig. 7.5 Ball on a plane. Figures **b-f** depict the time evolutions of the rolling torque τ_{rx} around the x-axis and of the Coulomb-Contensou friction forces λ_{ccz} and λ_{ccy} as well as the time evolutions of the angular velocities ω_z and ω_x

$\mu = 0.1$. The first ball is given an initial velocity $u_y = 1$ m/s in y-direction and an initial angular velocity $\omega_y = 20$ rad/s around the y-axis. The second ball is at rest. The contact between the first ball and the plane is denoted as contact A, the contact between the second ball and plane as contact B and the contact between the two balls as contact C. Contacts A and B are modeled with Coulomb-Contensou and rolling friction, contact C with spatial friction only. Figure 7.6b depicts the trajectories of the two balls in the x-y-plane. The discrete states $\hat{\sigma}$ of the discrete set-valued laws are shown in Fig. 7.6c. The occurring switching incidents are listed in Table 7.6. Note that due to the initial rotation around the y-axis, the first ball's movement is slightly curved at the beginning. At the switching point at $t = 0.198$s the

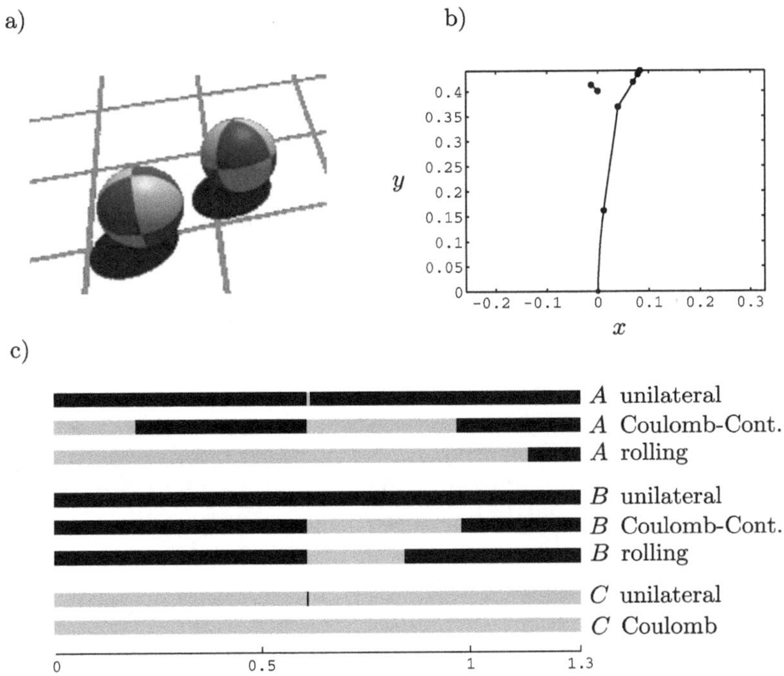

Fig. 7.6 Two balls on a plane. Figure **b** depicts the trajectories of the two blass in the plane, Fig. **c** indicates the discrete state $\hat{\sigma}$ of the discrete set-valued laws. The contacts between first ball and plane, second ball and plane and between the balls are referred as contacts A, B and C

Table 7.6 Switching incidents for two-ball-problem

time [s]	contact	set-valued law	action
0.198	A	Coulomb-Contensou	slip to stick
0.610	A	unilateral	closed to open
	B	Coulomb-Contensou	stick to slip
	B	rolling friction	constraint to impressed mode
	C	unilateral	open to closed
0.610	C	unilateral	closed to open
0.612	A	unilateral	open to close
	A	Coulomb-Contensou	slip
	A	rolling friction	impressed mode
0.843	B	rolling friction	impressed to constraint mode
0.968	A	Coulomb-Contensou	slip to stick
0.980	B	Coulomb-Contensou	slip to stick
1.14	A	rolling friction	impressed to constraint mode

Coulomb-Contensou law comes into stick state and the ball rolls straight on. At $t = 0.610$s the first ball collides with the second ball. The Coulomb-Contensou law of both contacts A and B switch into slip state. Due to the impactive friction force in contact C both balls obtain an angular velocity around the z-axis. In addition, the impulsive friction force causes the first ball to make a slight jump. At $t = 0.843$s the rolling friction law of contact C switches to constraint mode, i.e. the second ball rotates only around the vertical z-axis. At $t = 0.980$s the second ball comes to a rest. The first ball first switches at $t = 0.968$s into a pure rolling state, i.e. the Coulomb-Contensou law of contact A switches from slip to stick. The ball comes to a rest at $t = 1.14$s.

7.6.1.3 Rattleback

We consider bodies of ellipsoidal shape which are in contact with a plane. The positions $\mathbf{q}$ and the translational components of $\mathbf{u}$ are evaluated in the inertial system I, the rotational components of $\mathbf{u}$, i.e. Ω, in the body-fixed system B. Between the bodies and the plane acts a unilateral contact and Coulomb-Contensou friction.

The rattleback toy consists of an ellipsoid whose principal axes of inertia are slightly twisted against the geometric axes, see figure 7.7. The resulting deviatoric inertias cause the rattleback to turn its direction of rotation if started in the wrong direction. The kinetic energy of the initial rotation is first transferred to a waggling motion around the body-fixed x- and y-axis. Next, the kinetic energy is transferred back to the rotation around the body-fixed z-axis, but now in the opposite direction. The following parameters have been used for the simulation: principal geometric axes $a = 7.5$ cm and $b = 2$ cm, mass $m = 88$ g, moments of inertia $A = 0.0168 \cdot 10^{-3}$ kg m^2, $B = 0.1032 \cdot 10^{-3}$ kg m^2, $C = 0.1272 \cdot 10^{-3}$ kg m^2 and $F = -0.0157 \cdot 10^{-3}$ kg m^2. The unilateral contact's restitution coefficient is $\varepsilon_N = 0.5$. This choice reduces the drift into the plane due to the waggling which occurs at the turning point. The maximal Coulomb friction force is $0.2\,\lambda_N$, the maximal drilling torque $1.3 \cdot 10^{-5}$ m λ_N. The initial conditions have been chosen

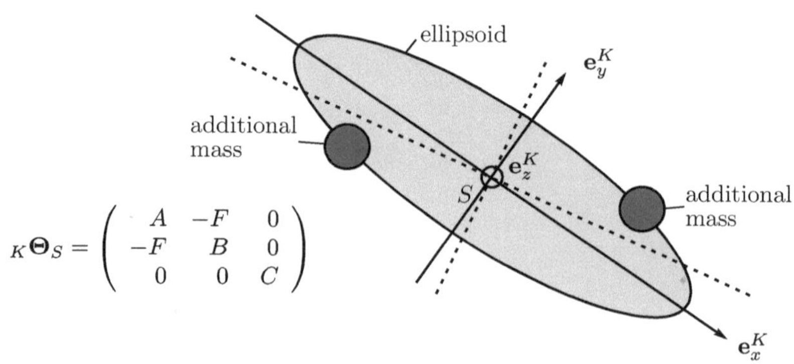

$$ {}_K\boldsymbol{\Theta}_S = \begin{pmatrix} A & -F & 0 \\ -F & B & 0 \\ 0 & 0 & C \end{pmatrix} $$

Fig. 7.7 The rattleback toy is an ellipsoid, whose principal axes of inertia are slightly twisted against the geometric axes

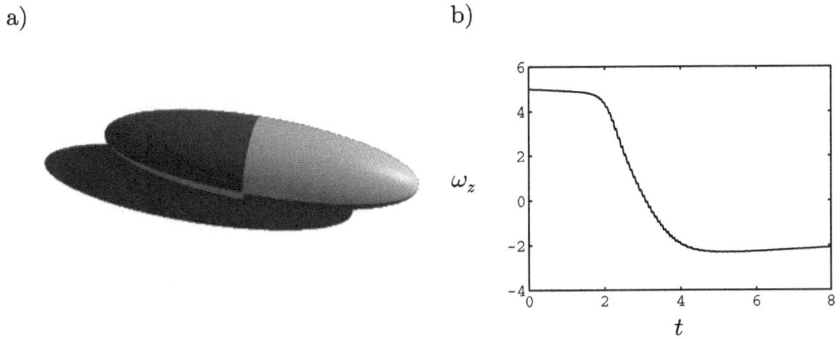

Fig. 7.8 Simulation results of the rattleback toy. The rattleback switches its turning direction. Figure **b** shows the time evolution of the angular velocity ω_z around the body-fixed z-axis, which stands initially perpendicular to the plane

as $\omega_y = 0.01$ rad/s and $\omega_z = 5$ rad/s, where the body-fixed z-axis stands initially perpendicular to the plane. Figure 7.8b depicts the time evolution of ω_z. The direction of rotation changes between $t = 2$s and $t = 4$s. In addition the drilling torque slightly decreases the absolute value of $|\omega_z|$.

7.6.1.4 Jumping Egg

The jumping egg is another example of a rotating ellipsoidal body on a plane [61]. In the initial position, the egg's body-fixed z-axis stands perpendicular to the plane. The egg is now given an initial spin around this body-fixed z-axis, which causes the egg to rise from the horizontal to the vertical, i.e. the center of mass rises against gravity. Within this motion, self-induced jumping can be observed. The following parameters have been used for the simulation: principal geometric axes $a = 3$ cm and $b = 2$ cm, mass $m = 13.3$ g, moments of inertia $A = 2.128 \cdot 10^{-5}$ kg m^2, and $B = C = 3.458 \cdot 10^{-5}$ kg m^2. The maximal Coulomb friction force is $0.2\,\lambda_N$, the

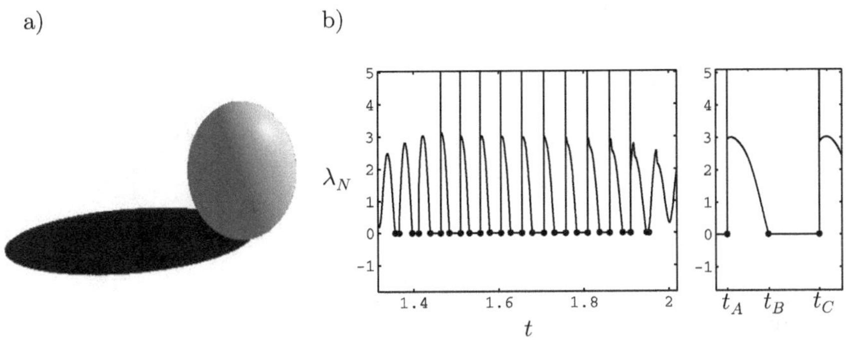

Fig. 7.9 Jumping egg. Figure **b** depicts the time evolution of the average contact force $\lambda_N = \frac{\hat{P}_N}{\Delta t}$

maximal drilling torque $1.3 \cdot 10^{-6}$ m λ_N. The initial conditions have been chosen as $\omega_x = \omega_y = 2$ rad/s and $\omega_z = 200$ rad/s. The average contact force $\lambda_N = \frac{\hat{P}_N}{\Delta t}$ in normal direction is depicted in Fig. 7.9b. The self-induced jumps correspond to $\lambda_N = 0$, i.e. the contact is closed between t_A and t_B and open between t_B and t_C, where $\Delta t = t_C - t_B$ varies between 10 ms to 25 ms.

7.6.1.5 Arbitrary Triangularized Bodies

Contact dynamics of arbitrary triangularized bodies is a very challenging task. The main problem arises from the need of a reasonable contact detection method, which must not only provide information whether the contact is closed or not but should also provide the contact's normal direction **n**. In addition, the contact detection should reduce the number of detected contact points in a reasonable sense. Consider for example two triangularized planes in contact. A common contact detection yields for this case an overwhelming number of dependent contacts, which makes the problem large and underdetermined. For the mechanical handling, only the contacts at the edges of the planes would be meaningful. A further problem is the varying number of contact points between two bodies. In order to keep the dimensions of the contact forces of the entire contact problem constant, a discrete set-valued law i can be defined which gathers a fixed number k of contacts, i.e. the discrete set-valued law consists of k unilateral and k friction elements. The number of contacts k is set equal to the maximal number of possible contacts expected between the two bodies. In general, this number k is larger than the number of detected contacts. In this case, the entries of the needless unilateral and friction elements in the generalized force directions $\mathbf{W}_i$ and in the force vector λ_i are filled with zeros. Due to the sparse representation of the matrices, such an approach does not affect the calculation time very much. Note that the $\text{prox}_{\hat{\mathscr{A}}_i}$ function of the discrete set-valued law i should solely process the necessary unilateral and friction elements. The dynamY package models triangularized bodies by vertices and triangles. The vertices are points in space, which are expressed in a body-fixed coordinate system. The triangles are defined by choosing these vertices as edge points. The dynamY contact

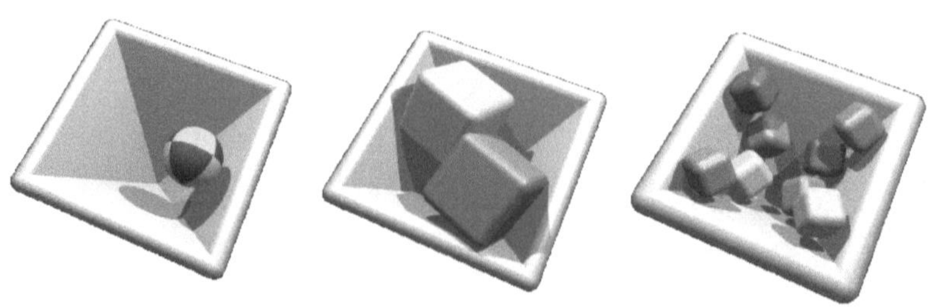

Fig. 7.10 Contact dynamics between a ball, boxes and funnel. The boxes and the funnel are represented by triangularized shapes

detection solves the contact problem by evaluating the following distances: all vertices of body A against all triangles of body B and vice versa, as well as all vertex connections (lines) of body A against all vertex connections of body B. The triangularized shape is given a thickness, which allows for detecting penetration. Planes are assumed to be modeled only by a small number of triangles, i.e. a box is modeled by 12 triangles. It may happen that one physical contact is detected twice due to the occurring penetration. Consider for example a vertex A which penetrates a triangle B. Due to the occurring non-vanishing penetration also a line which originates from the vertex A may be in contact with line of triangle B. For contacts which lie very close, it is therefore advised to consider only the contact with the maximal penetration. Figure 7.10 shows some snapshots obtained by dynamY .

7.6.2 Non-Common Set-Valued Laws

In the following, some non-common set-valued laws are discussed, i.e. set-valued laws for Stribeck friction, for pre-stressed springs and for remote action.

7.6.2.1 Stribeck Friction

The planar Stribeck friction law is depicted in Fig. 7.11. Note that the value of the sliding force λ_i depends on the relative velocity γ_i, i.e. the absolute value of the sliding force $|\lambda_i|$ decreases if the absolute value of the relative velocity $|\gamma_i|$ increases. However, such a friction law can be written as the sum of the classical set-valued friction law $-\gamma_i \in \mathscr{N}_{\mathscr{S}_p}(\lambda_{i1})$ and an additional single-valued Stribeck friction term

$$\lambda_{i2} = \left(\frac{1}{1+\delta|\gamma_i|} - 1\right) \cdot \text{sign}(\gamma_i), \tag{7.11}$$

where δ is an arbitrary positive parameter and $\text{sign}(\gamma_i)$ the sign of γ_i. Considering the dynamY environment, Stribeck friction is modeled by a derived nsc object which gathers the two forces λ_{i1} and λ_{i2} which are associated with the classical set-valued friction law and with the additional Stribeck friction term (7.11). Note that

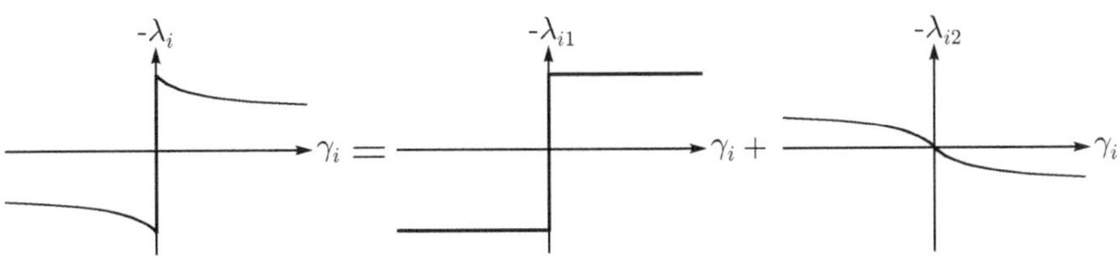

Fig. 7.11 Stribeck friction law as sum of the classical set-valued friction law $-\gamma_i \in \mathscr{N}_{\mathscr{S}_p}(\lambda_{i1})$ and a single-valued Stribeck term λ_{i2}

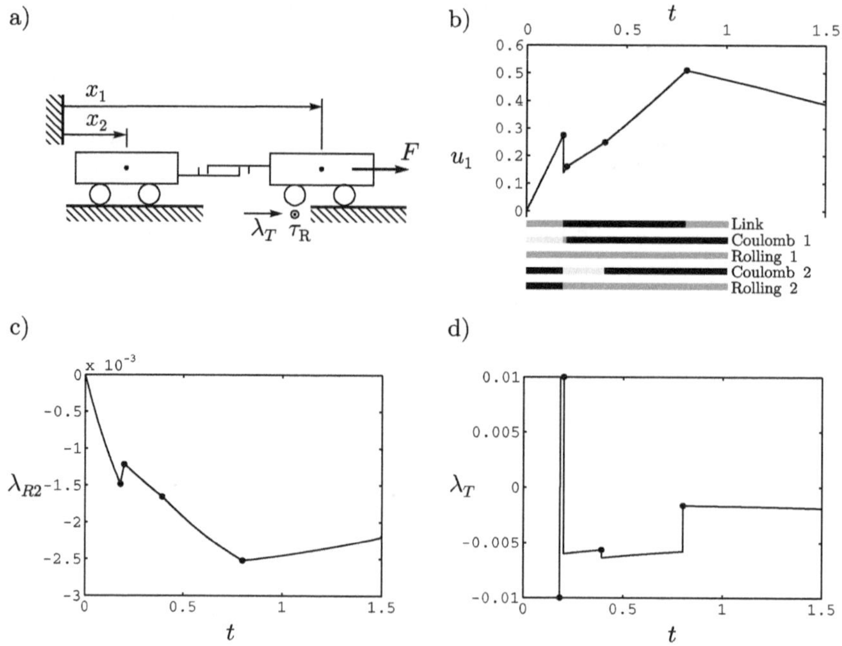

Fig. 7.12 Speeding-up of two waggons. The contact between the wheels and the the track is subjected to classical friction and a rolling Stribeck torque. Figure **b** depicts the time evolution of the first waggon's velocity u_1, Fig. **c** and **d** show the time evolution of the stribeck term $\lambda_{R2} = \tau_{R2}/r$ and of the tangential friction force λ_T . In addition, Fig. **b** depicts the discrete states $\hat{\sigma}$ of the discrete set-valued laws. *Black bars* correspond to constraint mode, *gray bars* to impressed mode. An additional distinction is made for friction, where *light gray bars* apply for sliding in positive direction and *dark gray bars* apply for sliding in negative direction

this term is evaluated by the nsc method nsc_initLawV, which obtains as input argument the global vector of all relative velocities γ. The classical set-valued friction law is implemented by the nsc_lawV method.

Take for instance the speeding up of a train, see Fig. 7.12. The contact between wheels and track is subjected to classical sliding friction λ_T and to a Stribeck rolling torque τ_R. The Stribeck rolling torque is represented by a force λ_R, which is scaled by the radius of the wheel r, i.e. $\tau_R = r\lambda_R$. The Stribeck rolling torque has a classical set-valued part $-\gamma_R \in \mathcal{N}_{\mathcal{S}_p}(\lambda_{R1})$ and a Stribeck part λ_{R2} according to (7.11). An external pulling force F acts on the first waggon. Each waggon is described by a translational coordinate x_i and two rotational coordinates φ_{i1} and φ_{i2} for the wheels. Due to Stribeck friction, the speeding up of the train requires less force if the waggons are connected loosely, compared to the case for which the buffers are at the limit. In the first case, each waggon starts to move individually, thus only the maximal Stribeck rolling resistance of a single waggon must be overcome. In Fig. 7.12 the speeding-up of two waggons is depicted. The initial conditions are chosen such that the buffers are loose and that the train is at rest. The following parameters have

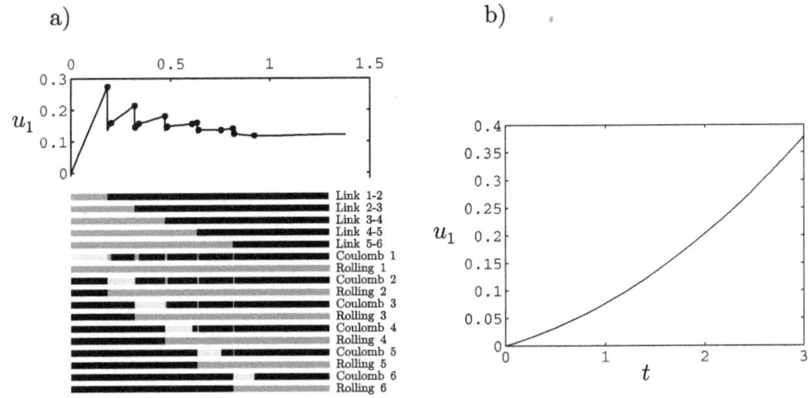

Fig. 7.13 Speeding-up of a train with 6 waggons. Figure **a** depicts the speeding up for a pulling force $F = 0.05$ N, where the links between the waggons are initially loose. In order to speed up a train with tensed links, the pulling force has to be increased to $F = 0.7$ N, see Fig. **b**

been chosen (toy train): mass of one waggon m=20 g, radius of wheels $r = 1$ cm, inertia of the wheels $5 \cdot 10^{-7}$ kg m^2, pulling force $F = 0.05$ N, maximal friction force $a_T = 0.01$ N, maximum rolling torque $\tau_{R,max} = r a_R = r 0.005$ N, Stribeck parameter $\delta = 2$ and restitution coefficient of the link between the waggons $\varepsilon = 0$. The inertias of the wheels are chosen quite large to underline some frictional effects.

Figures 7.12b,c and d depict the time evolution of the velocity $u_1 = \dot{x}_1$ of the first waggon, the Stribeck term λ_{R2} and the friction force λ_T. First, the wheels of the first waggon will be in sliding mode, where the sliding force accelerates the rotation of the wheels. After some time, the link between the waggons closes, which causes the velocity of the first waggon to decrease and the velocity of the second waggon to increase instantaneously. The wheels of the first waggon rotate now too fast, which causes the sliding force λ_T to change its direction and to slow down the rotation of the wheel. Note the associated peak in the Stribeck friction term λ_{R2}. On the other hand, the wheels of the second waggon accelerate in the other direction. After some time, all wheels are in pure rolling state. The friction resistance becomes minimal and the acceleration increases, which cause the friction force λ_T to increase and the Stribeck term λ_{R2} to decrease. Note that also in the pure rolling state the friction force λ_T is not constant due the coupling with the velocity-dependent Stribeck rolling model. At the time $t = 0.8$s the external force F is switched off and the train is slowed down by the rolling torque.

In the following, the speeding-up of a train with 6 waggons is investigated. The train will start to slide if the pulling force F is larger than the tangential friction resistance, i.e. $F > 2n\, a_T = 0.12$ N, where 2n denotes the number of wheels of the $n = 6$ waggons. The train starts rolling if $F > 2n\, a_R = 0.06$ N.

Figure 7.13a depicts the velocity u_1 of the first waggon, where the pulling force is $F = 0.05$ N, which is sufficiently large to speed up one waggon. The links between the waggons are initially loose. The waggons speed up one after each other. If the waggons are initially in tension, the pulling force must be increased to $F = 0.07$ N in order to speed up the train. The associated velocity u_1 of the first waggon is depicted in figure 7.13b. Note that the velocity u_1 increases nonlinearly due to the velocity-dependent Stribeck friction torque.

7.6.2.2 Pre-Stressed Spring

Next, a geometric set-valued force law which models a pre-stressed spring is discussed [41], see Fig. 7.14a. The element consists of two pre-stressed springs with tensioning force a_i and spring stiffness c_i. The tensioning is sustained by the two unilateral contacts. If the absolute value of the external force $|\lambda_i|$ exhibits the tensioning force a_i, then one of the two unilateral contact opens. Else, the pre-stressed spring keeps its reference length g_{0i}. The associated characteristic is depicted in Fig. 7.14b. As in the case of Stribeck friction, the characteristic can be divided in a set-valued part

$$-g_i \in \mathscr{N}_{\mathscr{S}_p}(\lambda_{i1}), \quad \mathscr{S}_p = [-a;a], \tag{7.12}$$

and in a single-valued term $\lambda_{i2} = -c\, g_i$. This distinction is not necessary anymore if the geometric set-valued force is expressed on velocity level,

$$g_i = 0 \Rightarrow -\gamma_i \in \mathscr{N}_{\mathscr{S}_p}(\lambda_i), \tag{7.13}$$

$$g_i \neq 0 \Rightarrow \lambda_i = -\big(a\,\mathrm{sign}(g) + c\,g\big). \tag{7.14}$$

Such a force law can be implemented in the dynamY environment by using the methods nsc_lawD for the case $g_i \neq 0$. Note that in a numeric sense g_i is always unequal to zero. Therefore, the user has to provide a tolerance δ in order to determine whether the inclusion (7.13) or the impressed mode (7.14) is applicable. Note that different to the unilateral contact, a pre-stressed spring element in impressed mode can not be neglected. The associated force $\lambda_i \neq 0$ must be considered separately in the same way as the vector of external and gyroscopic forces **h**.

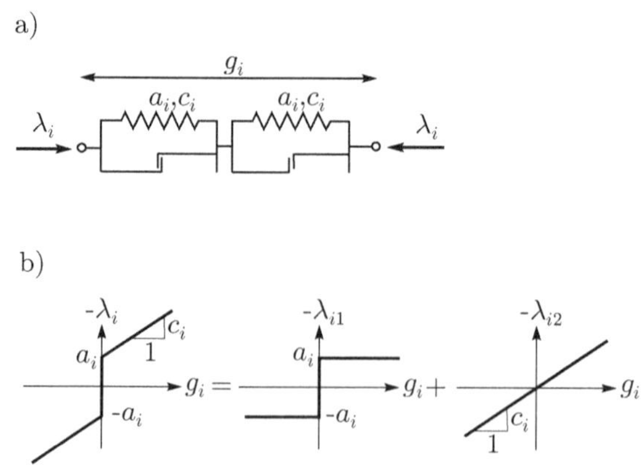

Fig. 7.14 Pre-stressed spring, model and characteristic [41]

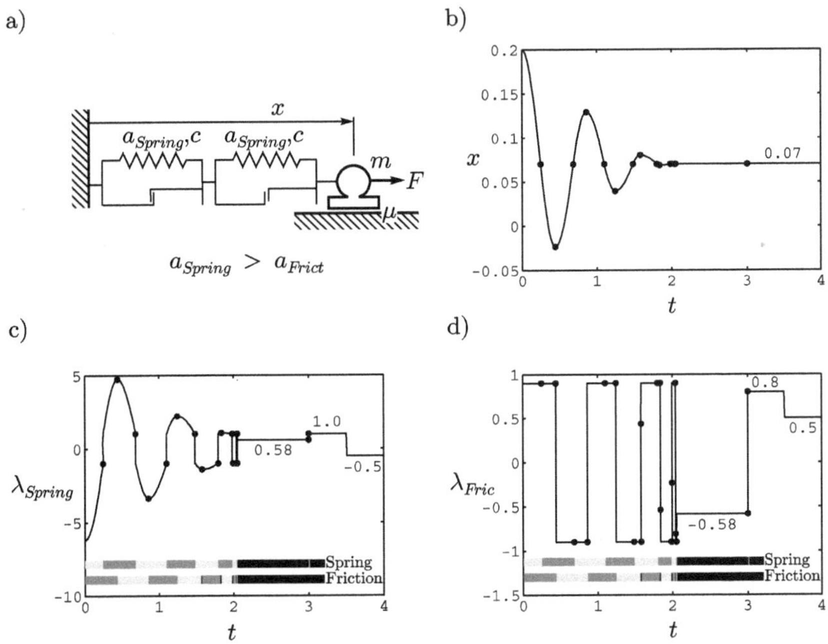

Fig. 7.15 Pre-stressed oscillator, where $a_{Fric} < a_{Spring}$. Figure **b** depicts the coordinate x, Fig. **c** the spring force λ_{Spring} and Fig. **d** the friction force λ_{Fric}. In addition, the discrete state $\hat{\sigma}$ of the discrete set-valued law is depicted. *Black bars* correspond to constraint mode, *gray bars* to impressed mode. An additional distinction is made for friction, where *light gray bars* apply for sliding in positive direction and *dark gray bars* apply for sliding in negative direction

As an example, consider a point mass m which is connected to the inertial surrounding by a pre-stressed spring, see Fig. 7.15a. Furthermore, the movement of the point mass is subjected to friction. An external force F acts on the point mass. The position and the velocity of the point mass are described by the coordinates x and $u = \dot{x}$, respectively. The tensioning force of the spring is denoted by a_{Spring}, the maximal friction force by a_{Fric}. It is of importance whether $a_{Fric} < a_{Spring}$ or $a_{Fric} > a_{Spring}$. In the first case, the movement of the system will always end up in the reference position $x = g_0$ if $F = 0$. In the second case, the end position will differ from the reference position, i.e. $x \neq g_0$. The following parameters have been used for the simulation: point mass $m = 1$ kg, spring stiffness $c = 40$ N/m, tensioning force $a_{Spring} = 1$ N, reference length $g_0 = 7$ cm, initial position $x_0 = 20$ cm, initial velocity $u_0 = 0$ m/s and tolerance $\delta = 0.5$ mm. The external force $F = 1.8$ N acts only for $3\text{s} \leq t \leq 3.5\text{s}$. Two simulation runs have been performed. For the first run, the maximal friction force $a_{Fric} = 0.9$ N was chosen less than the tensioning force a_{Spring}. The results are depicted in Figs. 7.15b-d. Figure 7.15b depicts the time evolution of the coordinate x. The oscillation ends up in the reference position $x = g_0 = 0.07$ m. Note the switching incidents. The friction force changes its direction at the peaks of $x(t)$, the pre-stressed spring switches at $x(t) = g_0$. After 3s, the external force

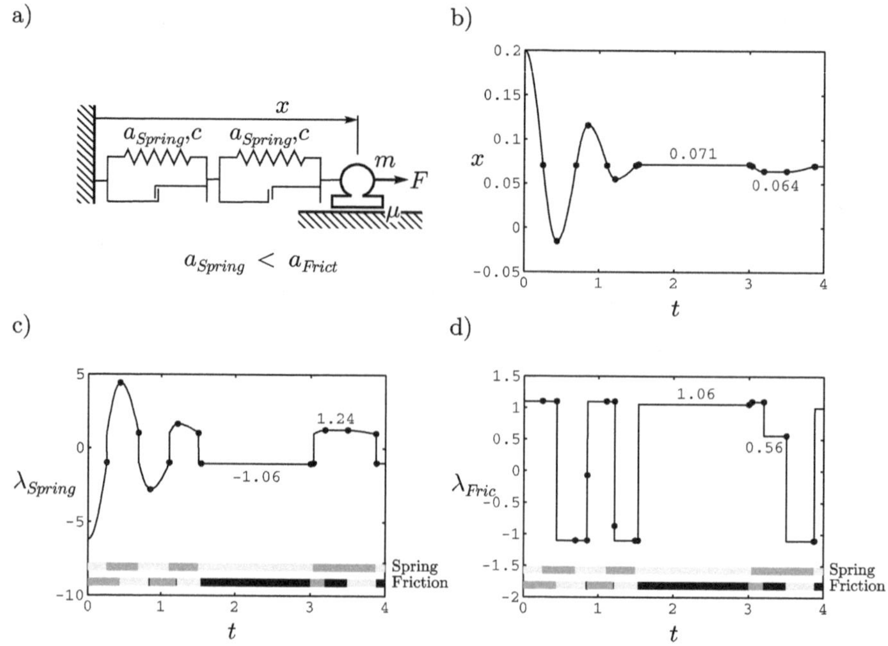

Fig. 7.16 Pre-stressed oscillator, where $a_{Fric} > a_{Spring}$. Figure **b** depicts the coordinate x, Fig. **c** the spring force λ_{Spring} and Fig. **d** the friction force λ_{Fric}

$F = 1.8$ N is switched on. Since this force is smaller than the resistance due to friction and tensioning, i.e. $a_{Fric} + a_{Spring}$, the coordinate x is unaffected. In Figs. 7.15c and d the time evolution of the spring force λ_{Spring} and of the friction force λ_{Fric} are depicted. Note especially that for $t > 2.055$s the system is at rest, i.e. the friction element sticks and the pre-stressed spring has its reference length g_0. The sum of both forces λ_{Spring} and λ_{Fric} corresponds to the external force F. Note that the individual values of the two forces are not uniquely defined, only the sum of the two. Some unexpected switching incidents occur at $t = 1.56$s, $t = 1.84$s and $t = 2$s. At these points the pre-stressed spring passes its reference length g_0, which causes the friction element to stick for a very short time. This behaviour must most probably be seen in connection with numerical problems related to the tolerance δ. Also at $t = 3$s an unexpected short opening of the pre-stressed spring occurs. Nevertheless, these unexpected switching incidents do rarely affect the motion of the system.

Figures 7.16b-d depict the results of a simulation with $a_{Fric} = 1.1$ N being larger than a_{Spring}. As a consequence, the oscillations do not end up in the reference position $x = g_0$, see Fig. 7.16b. Figures 7.16c and d depict the time evolution of the spring force λ_{Spring} and the friction force λ_{Fric}. Considering the stationary position at $x = 0.071$ m, the pre-stressed spring element is still in impressed mode, i.e. the absolute value of the spring force $|\lambda_{Spring}|$ is larger than the tensioning force a_{Spring}. Note especially the reaction of the pre-stressed spring to the switching on of the external force $F = 1.8$ N at $t = 3$s. As the pre-stressed spring is in impressed mode, it can not compensate the additional force without a displacement. The same applies at $t = 3.5$s when the force is again switched off.

7.6.2.3 Remote Action

By now impacts have been described by Newton's extended impact law in inequality form (3.36). Note that this impact law does only describe impacts locally. If impacts can not be described by Newton's extended impact law we speak of remote action or long distance effect. In this case, wave effects and contact stiffnesses play an important role in the evaluation of the impact behaviour. In the following, we discuss how remote action can be modeled. When using Newton's extended impact law (3.36) in standard form, i.e. ε is a diagonal matrix, then remote action can not occur. This lack can be fixed by using a so-called Frémond matrix ε, which is a full matrix of restitution coefficients. Doing so, remote action can be modeled in a straightforward way. Take for instance Newton's cradle. Within the dynamY package we define a new derived nsc class which defines a Newton's cradle element which acts between three balls. Figure 7.17 depicts the motion of a Newton's cradle where $\varepsilon_{11} = 0$, $\varepsilon_{12} = \varepsilon_{21} = 1$ and $\varepsilon_{22} = 0$. One clearly observes that the pre-impact velocity of the first ball u_1 is transferred to the third ball's post-impact velocity u_3. The velocity of the second ball is not affected and remains zero, i.e. $u_2 = 0$. Specific Frémond matrices ε which were obtained by experiments are given by Payr et al. [71].

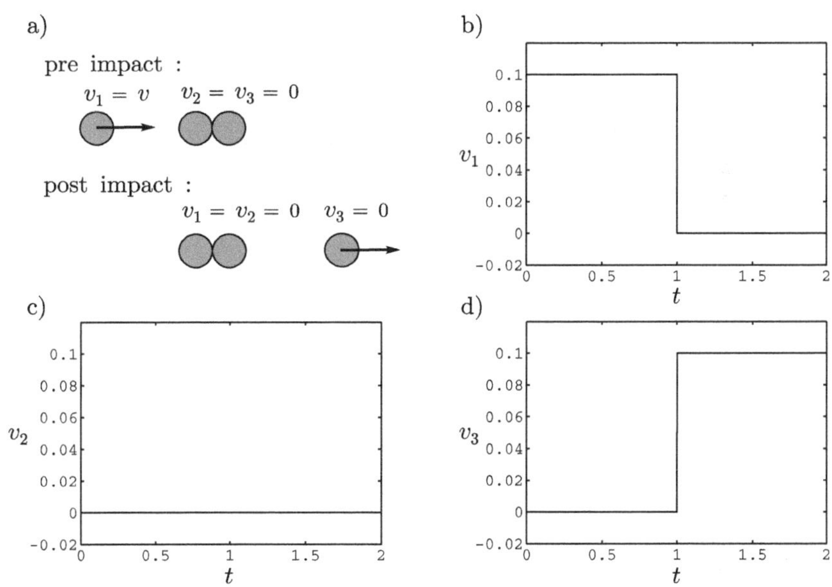

Fig. 7.17 Remote action for Newton's cradle. Figures **b**, **c** and **d** depict the velocities of the three balls for $\varepsilon_{11} = 0$, $\varepsilon_{12} = \varepsilon_{21} = 1$ and $\varepsilon_{22} = 0$

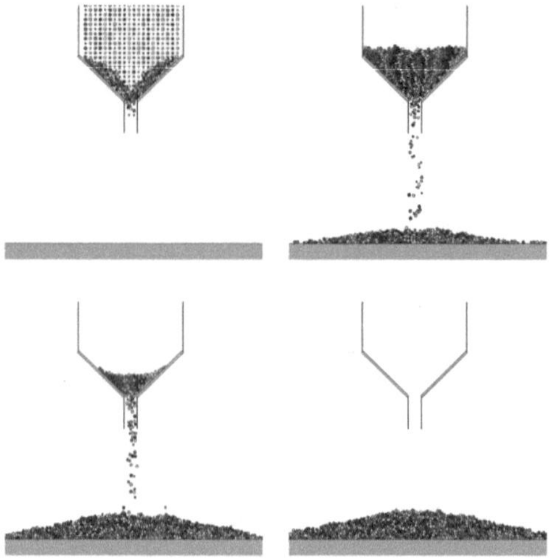

Fig. 7.18 Thousand balls falling in a funnel [93]

7.6.3 Granular Media

Time-stepping methods are very well suited for the simulation of granular material, as they do not aim to resolve all occurring switching points. Figure 7.18 shows some snapshots of 1000 balls which are thrown into a funnel. About half a million frictional unilateral contacts act between all the balls. Another example of granular application is the mixer shown in Fig. 7.19. In this case, thousand balls are mixed by a mixing tool which makes a sinusoidal movement. Note that such applications require quite long computation time. Considering the mixer, a 10s simulation with step size 0.001s requires about 2 days of computation time on a common desktop PC. Note that dynamY is not optimized for such computations, i.e. the software package aims more at generality than at speed optimization for special kinds of problems.

Fig. 7.19 Mixing of thousand balls [93, 94]

7.6.4 Motorbike

This example deals with the wheely of a motorbike. If a motorbike accelerates too fast, then the front wheel will separate from the ground and the bike will perform a wheely. The example aims at demonstrating that also more complex systems can be simulated within the dynamY environment. The model of the motorbike consist of 7 rigid bodies, which are connected by 6 rotational joints, one linear joint, two limit stops, 2 spring-dashpot elements and a module which models the chain. Furthermore, there are two frictional unilateral contacts between the wheels and the

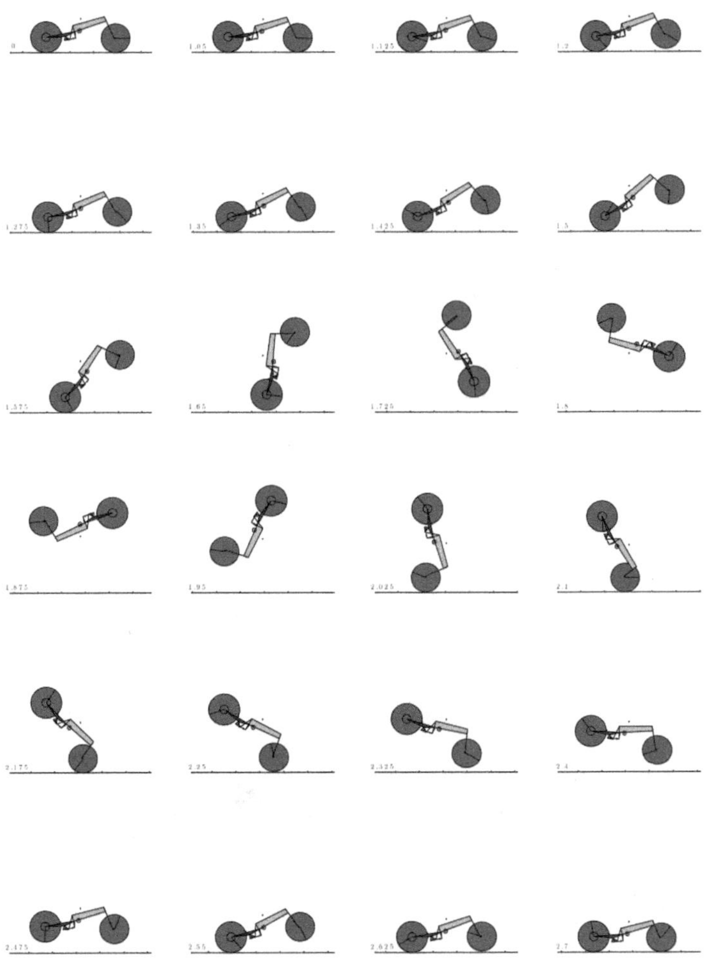

Fig. 7.20 Wheely of a motorbike

ground. The friction coefficient was chosen large in order to enable large accelerations. For the chosen typical parametrization, the motorbike makes a wheely at a critical acceleration of about 10 m^2/s (0 to 100 km/h in 2.7s). The critical driving torque is at about 320Nm. In Fig. 7.20 some snapshots of a simulation are shown. The driving torque was chosen equal to 400 Nm, where the torque applies only for $1s \leq t \leq 1.705s$. For such a choice, the motorbike performs an astonishing somersault. The motorbike was modeled in dynamY with the full set of coordinates, i.e. 21 degrees of freedom for the 7 rigid bodies. Objects derived from the nsc class were defined to model the unilateral contacts, limit stops, frictional elements and bilateral constraints. In order to prevent drift in the bilateral constraints, a constraint stabilization term $g_M/\Delta t$ according to Sect. 7.3 was used. Furthermore, gef objects were defined which model the spring-damper elements as well as the chain and the driving torque.

7.6.5 Elevator

Modeling elevators on long cables poses several interesting questions. When the cable of the elevator is wound on the reel, its strain is conserved until the cable is again released. As the loading might change in the meantime, oscillations might occur when releasing the cable. Consider the cable of an elevator which is wound on a reel. The cable is modeled as an elastic body, thus for each different loading the cable has another strain and tension. Due to friction, an infinitesimal cable element which is already deeply wound on the reel keeps its original strain and tension even when the loading of the elevator changes, i.e. the strain and the tension of the infinitesimal cable element is not adapted to the new loading until the element is released from the reel. Consider the following illustrating example. The elevator which is subjected to a loading F_1 begins to ascend. At some time the loading changes, i.e. the new loading becomes F_2 with $F_1 > F_2$. This change does not affect the infinitesimal cable elements which are already winded on the reel. Again, the elevator ascends. Then, the loading changes to F_3 with $F_1 > F_3 > F_2$, and the elevator starts to descend. We refer to all infinitesimal cable elements which have been wound under the loading F_1 as elements A and all infinitesimal cable elements which have been wound under the loading F_2 as elements B. When unwinding the rope, the elements B are stretched because $F_3 > F_2$. The elevator moves down faster than one would expect by the unwinding velocity. At some time the infinitesimal elements A are unwound. As $F_3 < F_1$, these elements contract themselves when released from the reel, and the elevator descends slower compared to the unwinding velocity.

The elevator was modeled in dynamY by a one-dimensional cable which was discretized using finite differences, i.e. the cable was modeled by n nodes which are connected by springs and dashpots, see Fig. 7.21. In the unstressed reference position, the cable has length L, which yields the distance between the nodes to be $\Delta x = L/n$. The scalar A denotes the cross-sectional area of A, the scalar E the modulus of elasticity, D a damping parameter and ρ the density of the cable. Each node i has an associated displacement q_i and velocity u_i, where the displacements describe the elongation with respect to the unstressed reference position, see Fig. 7.21b. The first node was given a velocity which is associated with the angular velocity of the reel, i.e. the cable is not winded on the reel but simply moves upwards (or downwards). If a node reaches a certain height, then a friction element is switched on which enforces the velocity of the node to be equal to the velocity of the first node. If the node falls below this critical height, then the associated friction element is switched off. The maximal friction force can be defined in various ways, for example it might be written as a function of the tension between two successive nodes. A more simple approach which was used in dynamY does not allow for sliding on the reel and uses a friction element with a "large" user-defined maximal friction force.

A challenging task when simulating the movement of the elevator is the wave propagation in the cable. In order to obtain a stable wave propagation, the numerical wave propagation velocity $v_{\text{num}} = \Delta x/\Delta t$ must be larger than the physical wave propagation velocity, which is around 5500 m/s for steel. The distances between the nodes Δx should be chosen quite small to account for a continuous unwinding. As a consequence of these requirements, the step size Δt must be chosen very small.

The following parameters have been chosen for the cable: Cross-sectional area $A = 7.8 \cdot 10^{-5}$ m^2, unstressed length $L = 20$ m, density $\rho = 7900$ kg/m^3, modulus of elasticity $E = 210 \cdot 10^9$ Pa and damping parameter $D = 2.5 \cdot 10^7$ Pa. The cable was discretized by $n = 100$ elements which yields $\Delta x = 0.2$ m. The time step size was chosen $1 \cdot 10^{-6}$s. The elevator cable is not subjected to gravity. The parameters

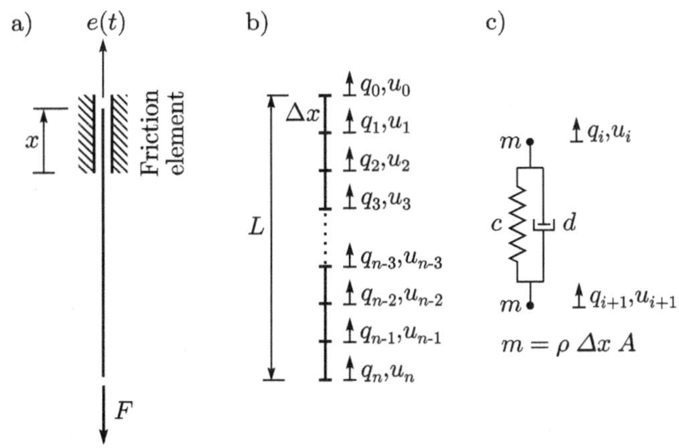

Fig. 7.21 Model of the elevator. Figure **a** depicts the model of the elevator which consists of a cable which can move up or downwards. The winding of the cable on the reel is modeled by a friction element which acts only on the parts of the cable which reach a certain height. The upper end of the cable is given a user defined velocity, the lower end of the cable is subjected to a loading F. Figure **b** depicts the finite difference discretization in the unstressed reference position. A finite difference discretization may be interpreted as nodes with mass $m = \rho \Delta x A$ which are connected by spring-dashpot elements with stiffness $c = EA/\Delta x$ and $d = DA/\Delta x$, see Fig. **c**

Table 7.7 Movement and loading of the elevator, see also Fig. 7.22a

time	time [s]	action
t_1	0.005	load $F_1 = 12$ kN, $x = 0$ m
t_2	0.2	elevator starts to ascend with $u_0 = 3$ m/s
t_3	2.3	elevator stops at $x = 6.3$ m
t_4	2.4	unloading, $F_2 = 0$ kN
t_5	2.7	elevator starts to ascend with $u_0 = 3$ m/s
t_6	4.8	elevator stops at $x = 12.6$ m
t_7	4.9	load $F_3 = 2$ kN
t_8	5.2	elevator starts to descend with $u_0 = -3$ m/s
t_9	9.4	elevator passes height at which F_1 has been unloaded
t_{10}	9.4	elevator stops at $x = 0$ m
t_{11}	9.5	unloading

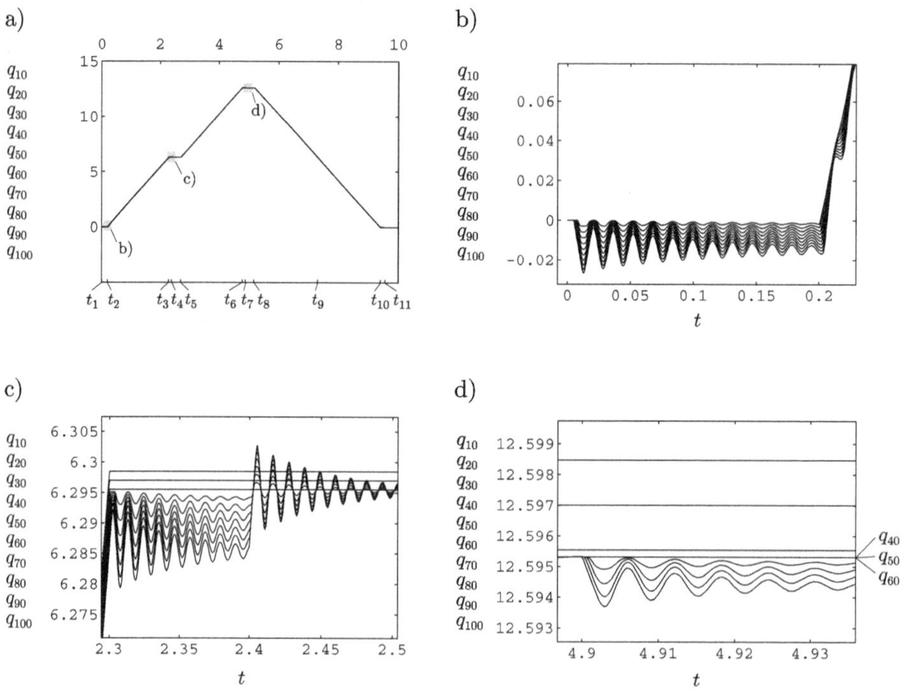

Fig. 7.22 Overall movement of the elevator according to Table 7.7

which describe the movement of the elevator and it's loading are given in Table 7.7. Figure 7.22 depicts the time evolution of the displacements $\mathbf{q}$. Note that only each 10-th coordinate q_i is plotted. Figure 7.22a depicts the overall movement according to Table 7.7. Figures 7.22b, 7.22c and 7.22d zoom into some interesting parts of Fig. 7.22a. In Fig. 7.22b the load F_1 is applied, and the elevator starts to ascend. Figure 7.22c depicts the situation at the first stop at $t = 2.3$s. The oscillations between $t = 2.3$s and $t = 2.4$s result from the stopping of the elevator, the further oscillations come from the unloading at $t = 2.4$s. Note that at the first stop the elevator

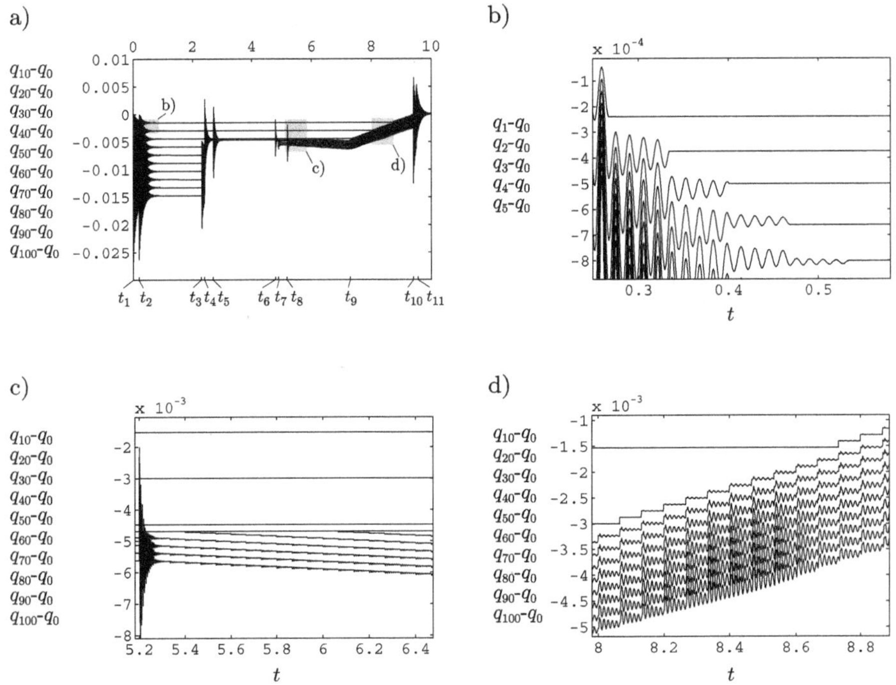

Fig. 7.23 Overall relative movement $q_i - q_0$ of the elevator nodes

has already reached a height of $x = 6.3$ m, thus only nodes with index $i \geq 32$ can oscillate freely. Due to the unloading, all these coordinates obtain an equal value. Figure 7.22d shows the loading of the elevator at $t = 4.9$s. The elevator has reached by now a height $x = 12.6$ m, thus only the nodes with index $i \geq 64$ can adapt to the new loading F_3. Figures 7.23a-d depict the relative positions $q_i - q_0$ of the nodes i with respect to the first node $i = 0$. In Fig. 7.23b it is shown how the elevator starts to ascend at $t_2 = 0.2$s. The relative movement $q_i - q_0$ is frozen when the node i reaches the friction element, i.e. the oscillation is cut off. Figure 7.23c shows the release of the nodes $i = 32 - 63$ when the elevator starts to descend at $t_8 = 5.2$s. Note that the strain and the tension between these nodes were frozen under the loading $F_2 < F_3$, thus the released cable is stretched. The opposite effect can be observed when releasing the nodes $1 - 31$, which is depicted in Fig. 7.23d. Note that the cable is not continuously unwound, i.e. the cable is released "node by node", which causes additional oscillations.

Chapter 8
Summary

During the last years, the interest in modeling mechanical systems with impacts and friction has increased enormously. New mathematical models have been developed, which account for the structure of such mechanical systems. It has to be distinguished between two main classes of models, the regularized models and the set-valued models. While the regularized models must be seen as an extension of classical smooth mechanics, the set-valued models use complete new formulations which *incorporate* classical mechanics. Set-valued models lead to non-smooth formulations, which have much less input parameters as regularized models and which are especially suited to understand the mechanisms of impacts and friction. In this book, the time integration of mechanical systems which consist of rigid bodies which can interact by set-valued laws has been discussed. The contacts and the friction elements have been modeled in a set-valued and thus non-smooth way. The set-valued force laws have been written as inclusions in a very general way which accounts for the various types of non-smooth interactions, like for example unilateral contacts, spatial friction, Coulomb-Contensou friction etc. It was shown how inclusion problems can be solved using an augmented Lagrangian approach. Time-stepping schemes have been discussed, and some extensions have been proposed which allow for a higher integration order and a step size adjustment. Special attention has been paid to keep all formalisms short and simple. We tried to give the various existing approaches on how to discretize a non-smooth system and on how to solve the resulting inclusion problems a common structure. As already mentioned, non-smooth mechanics includes also classical mechanics as most simple case. This can be verified in the modeling (inclusions reduce to equations), in the discretization (time-stepping schemes reduce to DAE integrators) but also in the solution process of the resulting inclusion problems (JORprox/SORprox reduce to classical JOR/SOR schemes).

In Chap. 3 the non-smooth modeling of a mechanical system has been discussed. It has been shown how unilateral contacts, friction and other non-smooth interactions can be described by set-valued force laws. These set-valued force laws have been expressed as normal cone inclusions, which yields a very general formulation which is

C. Studer: Numerics of Unilateral Contacts and Friction, LNACM 47, pp. 161–164.
springerlink.com

applicable for a broad class of non-smooth interactions. An important characteristic of the set-valued force laws and the associated normal cone inclusions is the convex set $\mathscr{C}_i$, which defines the set of admissible forces λ_i. If the force λ_i is in the interior of the set $\mathscr{C}_i$, then a constraint equation is enforced. The forces λ_i are set-valued and the law is in constraint mode. If the force λ_i is at the boundary of $\mathscr{C}_i$, then the system behaves in a way that the dissipation becomes maximal. In this case the set-valued force law is assigned to have impressed mode. Choosing $\mathscr{C}_i = \mathbb{R}$ yields the bilateral constraints, i.e. these constraints provide the most simple set-valued force law. The set-valued force laws have been connected to the equations of motion by a Lagrange I formulation. Non-smooth systems do not request the time evolution of the states $\mathbf{q}$ and $\mathbf{u}$ to be smooth, i.e. the velocities $\mathbf{u}$ might even jump at certain time instances. At such impacts, the accelerations and the forces λ are not defined anymore. Therefore, impact equations and impacts laws have been stated in order to account for the impulsive interactions. In order to obtain a comprehensive description of the mechanical system for both the impulsive and non-impulsive case, equalities of measures and associated set-valued laws have been formulated. The equality of measures and the set-valued laws reduce to the classical equations of motion and to the set-valued force laws if no impulsive event takes place. At an impact, the equalities of measures and the associated set-valued laws read as the impact equations and the impact laws. In this sense, the description of the mechanical system by the equality of measures and by set-valued laws is very general and incorporates *as special case* the whole classical dynamics subjected to bilateral constraints.

As already mentioned, set-valued laws can be represented by normal cone inclusions. In Chap. 4 the solution of such inclusion problems has been discussed in detail. The augmented Lagrangian approach, which aims at solving a constrained optimization problem, has been presented. Applying the approach to the dual principle of least constraints allows for transforming the inclusions which describe the non-smooth system into projective equations. These equations can be solved by iterating the forces λ in a Jacobi or Gauss-Seidel way. The iterative procedure corresponds to solving an underlying linear system by classical Jacobi or Gauss-Seidel iteration and to add specific projections on the solution in order to incorporate the non-smoothness. In this sense, the classical dynamics subjected to bilateral constraints must again be seen as the most simple special case, i.e. the associated projections to the set of all real number $\mathbb{R}$ becomes obsolete. The projective equations have been formulated in a global and in a local sense. The strength of a local formulation is that Euclidian projections to local convex sets can be used, which makes the projections simple. On the other hand, an iterative solution procedure is required when doing so. At first view, a global projective equation enables a solution of the inclusion problem in one step. Unfortunately, the associated projection is so expensive that the local approach still remains preferable.

In Chap. 5 different time-stepping methods have been discussed. These methods discretize directly the equalities of measures and the associated set-valued laws. The methods are very robust and simple in application. Time-stepping methods reduce in the special case of constraint equations to discretizations of differential algebraic equations, which result from classical mechanical systems subjected to bilateral constraints. Thus, the whole classical theory fits in whit this concept. Most problems which occur when formulating time-stepping schemes, such as drift problems and ill-conditioning, have their source in the underlying DAE discretization. The discrete set-valued laws associated with a time-stepping scheme have either been formulated on velocity or displacement level. Formulations on velocity level are capable of both partially elastic impacts and non-impulsive movement. In addition, the treatment of both geometric and kinematic set-valued laws is possible within the same framework. The only occurring problem is the drift. Formulations on displacement level are not subjected to drift. Due to the nonlinearity of the gap functions, the projective equations are not piecewise linear anymore. Therefore, formulations on displacement level are often linearized, which yields the discrete formulations to become only implicit in $\mathbf{u}_E$. This approach corresponds to a drift stabilized index-2 scheme, which is not capable of partially elastic impacts. A combination of set-valued force laws on velocity level with and without stabilization is possible as the stabilization term is evaluated explicitly. Set-valued laws which are subjected to drift might be stabilized, whereas set-valued laws which incorporate partially elastic impacts can be left in the unstabilized form. Note that the latter set-valued laws are not subjected to have pronounced drift problems.

In Chap. 6 it has been shown how the accuracy of Moreau's time-stepping scheme can be increased by using step-size adjustment and extrapolation. Switching points have been detected by observing the projection behaviour of the projective equations. This approach is very general and allows for a simple unified detection of switching points for arbitrary kind of set-valued laws. The switching points have been located by a regula falsi approach, and a minimal step size has been used to resolve the switching point. Doing so, it is still possible to manage accumulative switching points. In smooth parts of the motion, extrapolation has been used to increase the integration order. Moreau's time-stepping method has been taken as base integration scheme for extrapolation, i.e. different approximations have been computed by Moreau's scheme, which allows for the construction of higher order approximations. Only an uneven number of extrapolation substeps should be used in order to prevent numerical instabilities. By using a minimal step size to resolve switching points and by using extrapolation for the larger sized smooth time steps, a pseudo higher integration order has been obtained. Smooth time steps have been processed with classical higher integration order, while the errors at the switching points have been kept small by using a minimal step size.

In Chap. 7 the dynamY C++ software package, which can be used to simulate non-smooth systems, has been presented. The package allows for deriving classes which represent mechanical systems, which can be linked by non-smooth interactions. The presented examples underline the broad scope of different problems which can be approached by dynamY .

The modeling of non-smooth systems by inclusions is well established by now. Open questions remain in the field of multiple impacts, i.e. long distance effects and alternative impact laws. Considering the contact detection, contacts between triangularized bodies are still a challenging task. The problem consists in finding a reasonable selection of contact points which makes sense from a mechanical point of view. Much work on this topic has already been performed in the computer graphics society. The solution process of the resulting inclusion problem is well developed by now. However, this solution process is still the weakest point compared to the strong and robust time-stepping discretizations. Especially speed, non-unique solutions and reliability of convergence remain important questions. In this context also reduction techniques, which reduce the number of dependent set-valued laws, are of interest. Considering the time-stepping integration, the proposed extension to higher order integration by extrapolation and step size adjustment shows promising results and should be focussed further on. By now, time-stepping schemes are only available as research codes. However, it should be only a question of time until commercial software includes these elegant schemes in order to solve contact and friction problems in technical applications in a robust and simple way. Compared to the modeling and the integration of non-smooth systems, which are well understood nowadays, the control of non-smooth systems is still in the beginning of its development and will pose many interesting questions in the near future.

Appendix A
Glossary

This section provides an overview on abbreviations which are used in this work.

$\mathbf{M}$	mass matrix (positive definite and symmetric)
$\mathbf{h}$	vector of gyroscopic and external forces (including dashpot and spring forces)
$\mathbf{W}$	generalized force directions
ζ	inhomogeneity terms for relative contact velocities
$\hat{\zeta}$	inhomogeneity terms for relative contact accelerations
ε	matrix which contains restitution coefficients ε_i
$\mathbf{G}$	Delassus operator $\mathbf{G} = \mathbf{W}^\top \mathbf{M}^{-1} \mathbf{W}$, symmetric and at least positive semidefinite
$\mathbf{c}$	constant vector associated with the Delassus operator
$\mathbf{q}$	generalized coordinates
$\mathbf{u}$	generalized velocities
$\mathbf{u}^-, \mathbf{u}^+$	pre- and post-impact velocities
$\dot{\mathbf{u}}$	generalized accelerations
$\mathrm{d}\mathbf{u}$	measure for the generalized velocities
$\hat{\mathbf{u}}$	$= \Delta t\, \mathbf{u}$
$\mathbf{g}$	gap function
γ	relative velocities associated with all set-valued force laws
γ^-, γ^+	pre- and post-impact relative contact velocities
$\dot{\gamma}$	relative accelerations
$\mathrm{d}\gamma$	measure for the relative velocity
λ	forces associated with all set-valued force laws
Λ	impulsive forces
$\mathrm{d}\mathbf{P}$	measure for the percussions, $\mathrm{d}\mathbf{P} = \lambda\, \mathrm{d}t + (\Lambda^+ - \Lambda^-)\mathrm{d}\eta$
$\hat{\mathbf{P}}$	discrete percussions, $\int_{t_B}^{t_E} \mathrm{d}\mathbf{P}$
$\mathscr{P}$	$= \Delta t\, \hat{\mathbf{P}}$

C. Studer: Numerics of Unilateral Contacts and Friction, LNACM 47, pp. 165–167.
springerlink.com

$\mathscr{C}$	arbitrary convex set $\mathscr{C}$, set of admissible forces λ
$\mathscr{D}$	set of admissible impulsive forces Λ
$\mathscr{A}$	set of admissible percussion measures d**P**
$\hat{\mathscr{A}}$	set of admissible discrete percussions $\hat{\mathbf{P}}$
$\mathscr{S}_p$	set of admissible planar friction forces, $a =$ maximal friction force $\mu\lambda_j$
$\mathscr{S}_s$	set of admissible spatial friction forces, $a =$ maximal friction force $\mu\lambda_j$
$\mathscr{S}_a$	set of admissible anisotropic friction forces (ellipse-shaped)
$\mathscr{S}_{cc}$	set of admissible Coulomb-Contensou friction forces (ellipsoid-shaped)
$\mathbb{R}_0^+$	set of all positive real numbers, associated with unilateral contact
σ	state of the set-valued force law, $\sigma(t)$ state at time t
$\hat{\sigma}$	state of the discrete set-valued law, σ^s state which applies for entire time step s
$\mathbf{R}$	arbitrary positive definite symmetric matrix $\mathbf{R}$ associated with the norm $\lvert\lvert \cdot \rvert\rvert_{\mathrm{R}}$
r	scalar associated with a special choice of $\mathbf{R}$, associated with relaxation factor ω_h
ω_h	relaxation factor associated with row h
ρ	spectrum radius
$\mathscr{N}_{\mathscr{C}}$	normal cone associated with the convex set $\mathscr{C}$
$\Psi_{\mathscr{C}}$	indicator function associated with the convex set $\mathscr{C}$
$\Psi_{\mathscr{C}}^{\star}$	support function associated with the convex set $\mathscr{C}$
$\mathrm{prox}_{\mathscr{C}}^{\mathrm{R}}$	proximal point function to the convex set $\mathscr{C}$ with respect to $\lvert\lvert \cdot \rvert\rvert_{\mathrm{R}}$
$\mathrm{vdst}_{\mathscr{C}}^{\mathrm{R}}$	vector distance function
n	number of contacts (subproblems)
$\mathbf{x}_i$, λ_i	components of $\mathbf{x}$ associated with the i-th set-valued law, $i = 1 \ldots n$, for example λ_i is the force associated with the i-th set-valued law
m_i	dimension of subproblem, i.e. $\mathbf{x}_i \in \mathbb{R}^{m_i}$, for example: unilateral contact i, $m_i = 1$, spatial friction element i, $m_i = 2$
m	dimension of entire problem, i.e. $\mathbf{x} \in \mathbb{R}^m$
$\mathscr{C}_i$	local convex set $\mathscr{C}_i$, $\mathscr{C} = \mathscr{C}_1 \times \ldots \mathscr{C}_n$
$\mathbf{R}_i$	local matrix $\mathbf{R}_i$, $\mathbf{R} = \mathbf{R}_1 \oplus \ldots \mathbf{R}_n$
Δt	step size
$\mathbf{q}_B, \mathbf{u}_B$	states $\mathbf{q}$ and $\mathbf{u}$ at the beginning of the time interval
$\mathbf{q}_E, \mathbf{u}_E$	states $\mathbf{q}$ and $\mathbf{u}$ at the end of time the interval
$\mathbf{q}_M, \mathbf{u}_M$	states $\mathbf{q}$ and $\mathbf{u}$ at the midpoint of the time interval
$\mathbf{k}_D, \mathbf{K}_D$	vector (or matrix) evaluated for a specific parameter set associated with the index D, for example $\mathbf{q}_B = \mathbf{q}(t_B)$, $\mathbf{M}_B = \mathbf{M}(\mathbf{q}_B, t_B)$
s	index of time step
$\bar{\mathbf{g}}_E$, $\bar{\mathbf{h}}_E$	values of $\mathbf{g}$ and $\mathbf{h}$ obtained by linearization
$\mathbf{e}_L$	local integration error
$\mathbf{e}_G$	global integration error
p	order of integration scheme
K_L	bounded constant associated with the local integration error, $\max_{t_i} \lvert\lvert \mathbf{e}_L(t_i) \rvert\rvert \leq K_L \Delta t^{p+1}$
K_P	bounded constant associated with the propagation of the local integration error
K_G	bounded constant associated with the global integration error, $\lvert\lvert \mathbf{e}_G \rvert\rvert \leq K_G \Delta t^{p+1}$
$T_{i,j}$	approximation gained by extrapolation

Furthermore, the following terminology applies:

unilateral contact prevents penetration between bodies

frictional unilateral contact prevents penetration between bodies, where friction is present in the tangent contact plane, is modeled by two set-valued force laws, one for the unilateral behaviour and one for the frictional behaviour

set-valued force law force law for the non-impulsive case, describes for example an opening of a unilateral contact or a slip-stick transition of a friction element

set-valued law force law for both the impulsive and non-impulsive case, stated in terms of measures, describes for example an impact but also an opening of an associated unilateral contact

discrete set-valued law set-valued law in a discrete sense

forces λ forces associated with set-valued force laws, i.e. friction forces, contact forces etc.

relative velocities γ relative velocities associated with set-valued force laws

Appendix B
Projection to an Ellipse or to an Ellipsoid

In section 2.3.3 the solution procedure for proximal point functions to an ellipse- or ellipsoidal-shaped convex sets $\mathscr{S}_a$ and $\mathscr{S}_{cc}$ has been sketched, i.e. $\mathbf{x} = \text{prox}_{\mathscr{S}_a}(\xi)$ and $\mathbf{x} = \text{prox}_{\mathscr{S}_{cc}}(\xi)$. In this appendix some intermediate steps of the calculation are provided. It is assumed that ξ is not an element of the set $\mathscr{S}_a$ or $\mathscr{S}_{cc}$, respectively.

B.1 Projection to Ellipse Contour

In this section we give some intermediate calculation steps for the the projection of a point $\xi = (\xi_1\ \xi_2)^\top$ to an ellipse contour

$$e(\mathbf{x}) = \left(\frac{x_1}{c}\right)^2 + \left(\frac{x_2}{d}\right) - 1 = 0. \tag{B.1}$$

The gradient is

$$\nabla e(\mathbf{x}) = \begin{pmatrix} 2x_1/c^2 \\ 2x_2/d^2 \end{pmatrix}. \tag{B.2}$$

The forth order polynomial for α yields

$$\begin{aligned} 0 &= a_4\,\alpha^4 + a_3\,\alpha^3 + a_2\,\alpha^2 + a_1\,\alpha + a_0, \\ a_4 &= 16, \\ a_3 &= 16\,(c^2 + d^2), \\ a_2 &= 4\,(c^4 + 4c^2d^2 + d^4 - c^2\xi_1^2 - d^2\xi_2^2), \\ a_1 &= 4\,(c^4d^2 + c^2d^4 - c^2d^2\xi_1^2 - c^2d^2\xi_2^2), \\ a_0 &= c^4d^4 - c^2d^4\xi_1^2 - c^4d^2\xi_2^2, \end{aligned} \tag{B.3}$$

and the proximal point $\mathbf{x}$ is obtained by

C. Studer: Numerics of Unilateral Contacts and Friction, LNACM 47, pp. 169–170.
springerlink.com

$$x_1 = \frac{\xi_1}{1+\frac{2\alpha}{c^2}}, \tag{B.4}$$

$$x_2 = \frac{\xi_2}{1+\frac{2\alpha}{d^2}}. \tag{B.5}$$

B.2 Projection to Ellipsoid Contour

In this section we give some intermediate calculation steps for the the projection of a point $\xi = (\xi_1\, \xi_2\, \xi_3)^\top$ to an ellipsoid contour

$$e(\mathbf{x}) = \left(\frac{x_1}{c}\right)^2 + \left(\frac{x_2}{c}\right)^2 + \left(\frac{x_3}{d}\right)^2 - 1 = 0. \tag{B.6}$$

The gradient is

$$\nabla e(\mathbf{x}) = \begin{pmatrix} 2x_1/c^2 \\ 2x_2/c^2 \\ 2x_3/d^2 \end{pmatrix}. \tag{B.7}$$

The forth order polynomial for α yields

$$\begin{aligned}
0 &= a_4\,\alpha^4 + a_3\,\alpha^3 + a_2\,\alpha^2 + a_1\,\alpha + a_0, \qquad \text{(B.8)}\\
a_4 &= 16,\\
a_3 &= 16\,(c^2+d^2),\\
a_2 &= 4\,(c^4 + 4c^2d^2 + d^4 - c^2\xi_1^2 - c^2\xi_2^2 - d^2\xi_3^2),\\
a_1 &= 4\,(c^4d^2 + c^2d^4 - c^2d^2\xi_1^2 - c^2d^2\xi_2^2 - c^2d^2\xi_3^2),\\
a_0 &= c^4d^4 - c^2d^4\xi_1^2 - c^2d^4\xi_2^2 - c^4d^2\xi_3^2,
\end{aligned}$$

and the proximal point $\mathbf{x}$ is obtained by

$$x_1 = \frac{\xi_1}{1+\frac{2\alpha}{c^2}}, \tag{B.9}$$

$$x_2 = \frac{\xi_2}{1+\frac{2\alpha}{c^2}}, \tag{B.10}$$

$$x_3 = \frac{\xi_3}{1+\frac{2\alpha}{d^2}}. \tag{B.11}$$

References

1. Acary, V.: Contribution à la Modélisation Mécanique et Numérique des Edifices Macconnés. Ph.D. thesis, Ecole Supérieure de Mécanique de Marseille, Université d'Aix-Marseille II (2001)
2. Acary, V., Brogliato, B.: Numerical Time Integration of Higher Order Dynamical Systems with State Constraints. In: Proceedings ENOC 2005, Eindhoven (2005)
3. Acary, V., Jean, M.: Numerical Simulation of Monuments by the Contact Dynamics Method. In: DGEMN-LNEC-JRC (ed.) Monument 1998, Workshop on Seismic Perfomance of Monuments, pp. 69–78. Laboratório Nacional de engenharia Civil (LNEC), Lisboa (1998)
4. Acary, V., Jean, M.: Numerical Modeling of Three Dimensional Divided Structures by the Non Smooth Contact Dynamics Method. Application to Masonry Structures. In: Topping, B.H.V. (ed.) The Fifth International Conference on Computational Structures Technology 2000, pp. 211–222. Civil-Comp Press, Edimburgh (2000)
5. Acary, V., Pérignon, F.: Siconos: A Software Platform for Modeling, Simulation, Analysis and Control of Non Smooth Dynamical System. In: Proceedings of MATHMOD 2006, 5th Vienna Symposium on Mathematical Modelling. ARGESIM Verlag, Vienna (2006)
6. Aeberhard, U., Glocker, Ch.: Energy Considerations for Excited Perfect Collisions. In: Proceedings ENOC 2005, Eindhoven (2005)
7. Aeberhard, U., Payr, M., Glocker, Ch.: Theoretical and Experimental Treatment of Perfect Multi-Contact-Collisions. In: Proceedings ACMD 2006, Tokyo (2006)
8. Alart, P., Curnier, A.: A Mixed Formulation for Frictional Contact Problems Prone to Newton Like Solution Methods. Computer Methods in Applied Mechanics and Engineering 92(3), 353–375 (1991)
9. Anitescu, M., Hart, D.: A Constraint-Stabilized Time-Stepping Approach for Rigid Multibody Dynamics with Joints, Contact and Friction. Int. J. Numer. Meth. Engng. 60(14), 2335–2371 (2004)
10. Anitescu, M., Potra, F.A.: Formulating Dynamic Multi-Rigid-Body Contact Problems with Friction as Solvable Linear Complementarity Problems. Nonlinear Dynamics 14(3), 231–247 (1997)
11. Anitescu, M., Potra, F.A.: A Time-Stepping Method for Stiff Multibody Dynamics with Contact and Friction. Int. J. Numer. Meth. Engng. 55(7), 753–784 (2002)
12. Anitescu, M., Potra, F.A., Steward, D.E.: Time-Stepping for Three-Dimensional Rigid Body Dynamics. Computer Methods in Applied Mechanics and Engineering 177(3-4), 183–197 (1999)

13. Ballard, P., Basseville, S.: Existence and Uniqueness for Dynamical Unilateral Contact with Coulomb Friction. Mathematical Modelling and Numerical Analysis 39(1), 59–77 (2004)
14. Baraff, D.: Fast Contact Force Computation for Nonpenetrating Rigid Bodies. In: Computer Graphics Proceedings, Annual Conference Series 23–24 (1994)
15. Bertsekas, D.P.: Constrained Optimization and Lagrange Multiplier Methods. Academic Press, New York (1982)
16. Betsch, P.: The Discrete Null Space Method for Energy Consistent Integration of Constrained Mechanical Systems. Part I: Holonomic Constraints. Comput. Methods Appl. Mech. Engrg. 194(50-52), 5159–5190 (2005)
17. Betsch, P.: The Discrete Null Space Method for Energy Consistent Integration of Constrained Mechanical Systems. Part II: Multibody Dynamics. Int. J. Numer. Meth. Engrg. 67(4), 449–552 (2006)
18. Bohatier, C., Nouguier, C.: Contact Dynamics Between Tool and Granular Material. In: Proceedings Symphosium on Mathematical Theory of Networks and Systems (2000)
19. Botasso, C.L., Dopico, D., Trainelli, L.: Optimal Preconditioners for the Solution of Constrained Mechanical Systems in Index-3 From. In: Proceedings ACMD 2006, Tokyo (2006)
20. Brogliato, B.: Nonsmooth Impact Mechanics. Lecture Notes in Control and Information Sciences. Springer, Heidelberg (1996)
21. Brogliato, B.: Impacts in Mechanical Systems. Lecture Notes in Physics. Springer, Heidelberg (2000)
22. Brogliato, B., ten Dam, A.A., Paoli, L., Génot, F., Abadie, M.: Numerical Simulation of Finite Dimensional Multibody Nonsmooth Mechanical Systems. ASME Applied Mechanics Reviews 55(2), 107–150 (2002)
23. Cataldi, E.: Curve Squealing Mechanism of Railway Vehicles. Ph.D. thesis, ETH Zürich (2007)
24. Christensen, P.W., Klarbring, A., Pang, J.S., Stromberg, N.: Formulation and Comparison of Algorithms for Frictional Contact Problems. Int. J. Numer. Meth. Engng. 21(1), 145–173 (1998)
25. Cottle, R., Pang, W.: The Linear Complementarity Problem. Academic Press Inc., London (1992)
26. Deuflhard, P., Hairer, E., Zugck, J.: One-Step and Extrapolation Methods for Differential-Algebraic Systems. Numerische Mathematik 51(5), 501–516 (1987)
27. Dumont, Y., Paoli, L.: Simulations of Beam Vibrations Between Stops: Comparison of Several Numerical Approaches. In: Proceedings ENOC 2005, Eindhoven (2005)
28. Dzonou, R.: A Numerical Scheme for Finite Dimensional Frictionless Dynamics with Inelastic Contact and a General Inertia Operator. In: Proceedings ENOC 2005, Eindhoven (2005)
29. Ebrahimi, S., Eberhard, P.: Contact of Planar Deformable Bodies Using a Linear Complementarity Formulation. PAMM Proceedings in Applied Mathematics and Mechanics 5(1), 197–198 (2005)
30. Eich-Soellner, E., Führer, C.: Numerical Methods in Multibody Dynamics. B.G. Teubner, Stuttgart (1998)
31. Elstrodt, J.: Maß- und Integrationstheorie. Springer, Heidelberg (1996)
32. Förg, M., Geier, T., Neumann, L., Ulbrich, H.: r-Factor Strategies for the Augmented Lagrangian Approach in Multi-Body Contact Mechanics. In: Proceedings of III European Conference on Computational Mechanics, Lisbon (2006)
33. Förg, M., Pfeiffer, F., Ulbrich, H.: Simulation of Unilateral Constrained Systems with Many Bodies. Multibody System Dynamics 14(2), 137–154 (2005)

34. Förg, M., Ulbrich, H.: Analysis of Different Time-Integration Methods Applied to a Non-Smooth Industrial Problem. In: Proceedings ENOC 2005, Eindhoven (2005)
35. Frémond, M.: Rigid Bodies Collisions. Physical Letters A 204, 33–41 (1995)
36. Funk, K.: Simulation eindimensionaler Kontinua mit Unstetigkeiten, VDI-Fortschrittberichte Mechanik/Bruchmechanik, vol. 18/294. VDI-Verlag, Düsseldorf (2004)
37. Gavrea, B., Anitescu, M., Potra, A.: Convergence of a Class of Semi-Implicit Time-Stepping Schemes for Nonsmooth Rigid Multibody Dynamics. SIAM Journal of Optimization 19(2), 969–1001 (2008)
38. Geier, T., Foerg, M., Zander, R., Ulbrich, H., Pfeiffer, F., Brandsma, A., van der Velde, A.: Simulation of a Push Belt CVT Considering Uni- and Bilateral Constraints. ZAMM-Zeitschrift für angewandte Mathematik und Mechanik 86(10), 795–806 (2006)
39. Glocker, Ch.: Dynamik von Starrkörpersystemen mit Reibung und Stössen, VDI-Fortschrittberichte Mechanik/Bruchmechanik, vol. 18/182. VDI-Verlag, Düsseldorf (1995)
40. Glocker, Ch.: Formulation of Spatial Contact Situations in Rigid Multibody Systems. Comput. Methods Appl. Mech. Engrg. 177(3-4), 199–214 (1999)
41. Glocker, Ch.: Set-Valued Force Laws – Dynamics of Non-Smooth Systems. Lecture Notes in Applied Mechanics, vol. 1. Springer, Heidelberg (2001)
42. Glocker, Ch.: Simulation von harten Kontakten mit Reibung: Eine iterative Projektionsmethode. VDI-Berichte Nr. 1968: Schwingungen in Antrieben 2006 Tagung, Fulda. VDI-Verlag, Düsseldorf (2006)
43. Glocker, Ch.: Reduction Techniques for Distributed Set-Valued Force Laws. In: Baniotopoulo, C.C. (ed.) Proceedings of the 2nd International Conference on Nonsmooth/Nonconvex Mechanics with Applications Engineering, Ziti, Thessaloniki edn., Thessaloniki, Greece, July 2006, pp. 173–180 (2006)
44. Glocker, Ch., Studer, C.: Formulation and Preparation for Numerical Evaluation of Linear Complementarity Systems in Dynamics. Multibody System Dynamics 13(4), 447–463 (2005)
45. Hairer, E., Norsett, S.P., Wanner, G.: Solving Ordinary Differential Equations I. Springer, Heidelberg (2000)
46. Hairer, E., Norsett, S.P., Wanner, G.: Solving Ordinary Differential Equations II. Springer, Heidelberg (2000)
47. Jean, M.: The Non-Smooth Contact Dynamics Method. Computer Methods in Applied Mechanics and Engineering 177(3-4), 235–257 (1999)
48. Jean, M., Acary, V., Moniere, Y.: Non-Smooth Contact Dynamics Approach of Cohesive Materials. Phil. Trans. R. Soc. Lond. A 359(1789), 2497–2518 (2001)
49. Jourdan, F., Alart, P., Jean, M.: A Gauss-Seidel Like Algorithm to Solve Frictional Contact Problems. Comput. Methods Appl. Mech. Engrg. 155, 31–47 (1998)
50. Jourdan, F., Jean, M., Alart, P.: An Alternative Method Between Implicit and Explicit Schemes Devoted to Frictional Contact Problems in Deep Drawing Simulation. Journal of Materials Processing Technology 80(1), 257–262 (1998)
51. Kanzow, C.: Numerik linearer Gleichungssysteme. Springer, Berlin (2005)
52. Klarbring, A.: A Mathematical-Programming Approach to 3-Dimensional Contact Problems with Friction. Computer Methods in Applied Mechanics and Engineering 58(2), 175–200 (1986)
53. Lanier, J., Jean, M.: Experiments and Numerical Simulations with 2D Disks Assembly. Powder Technology 109(1-3), 206–221 (2000)
54. Le Saux, C., Leine, R.I., Glocker, Ch.: Dynamics of a Rolling Disk in the Presence of Dry Friction. J. Nonlinear Sci. 15(1), 27–61 (2005)

55. Leine, R.I.: On the Stability of Motion in Non-Smooth Meachnical Systems. Habilitation, ETH Zürich (2006)
56. Leine, R.I., Glocker, Ch.: A Set-Valued Force Law for Spatial Coulomb-Contensou Friction. European Journal of Mechanics A/Solids 22(2), 193–216 (2003)
57. Leine, R.I., Nijmeijer, H.: Dynamics and Bifurcations of Non-smooth Mechanical Systems. Lecture Notes in Applied and Computational Mechanics, vol. 18. Springer, Berlin (2004)
58. Lubich, C.: Linearly Implicit Extrapolation Methods for Differential-Algebraic-Systems. Numerische Mathematik 55(2), 197–211 (1989)
59. Lubich, C., Nowak, U., Pöhle, U., Engstler, C.: MEXX - Numerical Software for the Integration of Constrained Mechanical Multibody Systems, `http://na.uni-tuebingen.de/na/software/mexax/mexax.shtml`
60. Luenberger, D.G.: Linear and Nonlinear Programming. Addison-Wesley, Reading (1984)
61. Mitsui, K., Aihara, K., Terayama, C., Kobayashi, H., Shimomura, Y.: Can a Spinning Egg Really Jump? Proc. R. Soc. A 462(2074), 2897–2905 (2006)
62. Moeller, M., Glocker, Ch.: Non-Smooth Modelling of Electrical Systems Using the Flux Approach. Nonlinear Dynamics 50(1-2), 273–295 (2007)
63. Monteiro Marques, M.D.P.: Differential Inclusions in Nonsmooth Mechanical Problems. Shocks and Dry Friction. Birkhäuser, Basel (1993)
64. Moreau, J.J.: Bounded Variation in Time. In: Topics in Nonsmooth Mechanics, pp. 1–74. Birkhäuser Verlag, Basel (1988)
65. Moreau, J.J.: Unilateral Contact and Dry Friction in Finite Freedom Dynamics. In: Nonsmooth Mechanics and Applications. CISM Courses and Lectures, vol. 302. Springer, Wien (1988)
66. Moreau, J.J.: Some Numerical-Methods in Multibody Dynamics - Application to Granular Materials. European Journal of Mechanics A-Solids 13, 93–114 (1994)
67. Murty, K.G.: Linear Complementarity, Linear and Nonlinear Programming. Heldermann, Berlin (1988)
68. Paoli, L., Schatzman, M.: A Numerical Scheme for Impact Problems I: The One-Dimensional Case. SIAM Journal on Numerical Analysis 40(2), 702–733 (2002)
69. Paoli, L., Schatzman, M.: A Numerical Scheme for Impact Problems II: The Multi-Dimensional Case. SIAM Journal on Numerical Analysis 40(2), 734–768 (2002)
70. Payr, M., Glocker, Ch.: Oblique Frictional Impact of a Bar: Analysis and Comparison of Different Impact Laws. Nonlinear Dynamics 41(4), 361–383 (2005)
71. Payr, M., Glocker, Ch., Bösch, C.: Experimental Treatment of Multiple–Contact–Collisions. In: ENOC 2005 Proceedings (2005)
72. Pfeiffer, F., Foerg, M., Ulbrich, H.: Numerical Aspects of Non-Smooth Multibody Dynamics. Comput. Methods Appl. Mech. Engrg. 195(50-51), 6891–6908 (2006)
73. Pfeiffer, F., Glocker, Ch.: Multibody Dynamics with Unilateral Contacts. John Wiley and Sons, New York (1996)
74. Potra, F., Anitescu, M., Gavrea, B., Trinkle, J.: A Linearly Implicit Trapezoidal Method for Integrating Stiff Multibody Dynamics with Contacts, Joints and Friction. Int. J. Numer. Meth. Engng. 66(7), 1079–1124 (2006)
75. Powell, M.: A Method for Nonlinear Constraints in Minimization Problems. In: Optimization. Academic Press, London (1969)
76. Rockafellar, R.T.: Lagrange Multipliers and Variational Inequalities. Variational Inequalities and Complementarity problems
77. Rockafellar, R.T.: Convex Analysis. Princeton University Press, Princeton (1971)
78. Sayir, M., Dual, J., Kaufmann, S.: Ingenieurmechanik 1. B.G. Teubner (1997)

79. Schwarz, H.R.: Numerische Mathematik. B.G. Teubner, Stuttgart (1997)
80. Seifried, R.: Numerische und experimentelle Stossanalyse für Mehrkörpersysteme. Ph.D. thesis, University of Stuttgart, Stuttgart (2005)
81. SICONOS: European Project IST2001-37172, `http://siconos.inrialpes.fr/`, `http://siconos.gforge.inria.fr/index.html`
82. Simo, J.C., Laursen, T.: An Augmented Lagrangian Treatment of Contact Problems Involving Friction. Computers and Structures 42(1), 97–116 (1992)
83. Slavic, J., Boltezar, M.: Simulating Multibody Dynamics with Rough Contact Surfaces and Run-in Wear. Nonlinear Dynamics 45(3-4), 353–365 (2005)
84. Stewart, D.E., Trinkle, J.C.: An Implicit Time-Stepping Scheme for Rigid Body Dynamics with Inelastic Collisions and Coulomb Friction. Int. J. Numer. Methods Engineering 39(15), 2673–2691 (1996)
85. Stiegelmeyr, A.: Zur numerischen Berechnung strukturvarianter Mehrkörpersysteme, VDI-Fortschrittberichte Mechanik/Bruchmechanik, vol. 18/271. VDI-Verlag, Düsseldorf (2001)
86. Strehmel, K., Weiner, R.: Numerik gewöhnlicher Differentialgleichungen. B.G. Teubner, Stuttgart (1995)
87. Stronge, W.J., James, R., Ravani, B.: Oblique Impact with Friction and Tangential Compliance. Phil. Trans. R. Soc. Lond. A. 359(1789), 2447–2465 (2001)
88. Studer, C.: dynamY Software for Non-Smooth Systems (2008), `http://www.zfm.ethz.ch/dynamY`
89. Studer, C., Glocker, Ch.: A Time-Stepping Method for Non-Smooth Systems. PAMM-Proc. Appl. Math. Mech. 5, 217–218 (2005)
90. Studer, C., Glocker, Ch.: Simulation of Non-Smooth Mechanical Systems with Many Unilateral Constraints. In: Proceedings ENOC 2005, Eindhoven (2005)
91. Studer, C., Glocker, Ch.: Representation of Normal Cone Inclusion Problems in Dynamics via Nonlinear Equations. Archive of Applied Mechanics 76(5-6), 327–348 (2006)
92. Studer, C., Glocker, Ch.: Augmented Time Stepping by Step-Size Adjustment and Extrapolation. In: Proceedings IDETC/CIE-2007, Las Vegas (2007)
93. Studer, C., Glocker, Ch.: Solving Normal Cone Inclusion Problems in Contact Mechanics by Iterative Methods. Journal of System Design and Dynamics 1(3), 458–467 (2007)
94. Studer, C., Leine, R.I., Glocker, Ch.: Step-Size Adjustment and Extrapolation for Time Stepping Schemes in Non-Smooth Dynamics. Int. J. Numer. Meth. Engng. 76(11) (2008)
95. Transeth, A.A., Leine, R.I., Glocker, Ch., Pettersen, K.Y.: Non-Smooth 3D Modeling of a Snake Robot with Frictional Unilateral Constraints. In: Proceedings of the 2006 IEEE International Conference on Robotics and Biomimetics, Kunming, China (2006)
96. Trinkle, J.C., Pang, J.S., Sudarsky, S., Lo, G.: On Dynamics Mult-Rigid-Body Contact Problems with Coulomb Friction. ZAMM 77(4), 267–279 (1997)
97. Weinstein, R., Teran, J., Fedkiw, R.: Dynamic Simulation of Articulated Rigid Bodies with Contact and Collision. IEEE Transactions on Visualization and Computer Graphics 12(3), 365–374 (2006)
98. Welge-Lüssen, T., Glocker, Ch.: Experimental Passively Actuated Robot Arm for Non-Smooth Mechanics. In: Proceedings ENOC 2005, Eindhoven (2005)
99. Widmer, H.: Ein Simulationsmodell zur Untersuchung von piezoelektrischer Ultraschallantrieben. Ph.D. thesis, ETH Zürich (2008)
100. Yunt, K.: Trajectory Optimization of Structure-Variant Mechanical Systems. In: Proceedings of Int. Workshop on Variable Structure Systems 2006, Alghero, Italy, pp. 298–303 (2006)

List of Figures

Index

Kolibri 360 GmbH
Ferdinand-Jühlke-Straße 7
99095 Erfurt, Deutschland
produktsicherheit@kolibri360.de